Scott Rashid

Small Mountain Owls

Written and Illustrated by Scott Rashid

Schiffer Publishing Ltd

4880 Lower Valley Road · Atglen, PA 19310

Schiffer Books are available at special discounts for bulk purchases for sales promotions or premiums. Special editions, including personalized covers, corporate imprints, and excerpts can be created in large quantities for special needs. For more information contact the publisher:

Published by Schiffer Publishing Ltd.
4880 Lower Valley Road
Atglen, PA 19310
Phone: (610) 593-1777; Fax: (610) 593-2002
E-mail: Info@schifferbooks.com

For the largest selection of fine reference books on this and related subjects, please visit our web site at **www.schifferbooks.com**
We are always looking for people to write books on new and related subjects. If you have an idea for a book please contact us at the above address.

This book may be purchased from the publisher.
Include $5.00 for shipping.
Please try your bookstore first.
You may write for a free catalog.

In Europe, Schiffer books are distributed by
Bushwood Books
6 Marksbury Ave.
Kew Gardens
Surrey TW9 4JF England
Phone: 44 (0) 20 8392 8585; Fax: 44 (0) 20 8392 9876
E-mail: info@bushwoodbooks.co.uk
Website: www.bushwoodbooks.co.uk

Designed by RoS
Type set in Bernhard Modern BT/Humanist 521 BT

ISBN: 978-0-7643-3282-1
Printed in China

Dedication

To Dr. Ronald Ryder, Sigrid Ueblacker, and in memory of Jim Osterberg

Acknowledgments

This book would have never happened without the assistance of many individuals. I wish to thank Mr. Christian Cold, Wildlife Technician and Educator for the Bureau of Wildlife Management, Wisconsin Department of Natural Resources, in Ladysmith, Wisconsin, who was the one who introduced me to the practice of trapping and banding raptors.

Dr. Ronald Ryder, Professor Emeritus from Colorado State University in Fort Collins, Colorado, who inspired me to begin researching birds. Sigrid Ueblacker, founder and president of the Birds of Prey Foundation in Broomfield Colorado, who assisted me with all my rehabilitation questions, and Rick Spowart of the Colorado Division of Wildlife, who introduced me to both Ron and Sigrid.

I would also like to thank, Eric Adams, David Armstrong, John Barber, Alan Bell, Steve Bouricius, Burt Bowles, Chris and Linda Bieker, Michelle Blank, Laura MacAlister Brown, Heidi Buckman, Jeff Connor, Dan and Susan D'Amico, Hal Delzell, Fred and Tina Engelman, Nancy Gobris, Marie Hanabusa, Nancy Hause, Greg Hayward, Denver Holt, Eugene Jacobs, Wayne and Diana Johnston, Katherine McKeever, Ellen Holly Klaver, Brian Linkhart, Connie Kogler, Nick Komar, Kim Lankford, Mark Lindquist, Gary Mathews, Jeff Maugans, Jack and Lulie Melton, Steve Morello, Reagan Morgan, Jim and Bobbie Osterberg, Nathan Pieplow, Susan Rashid, John Rawinski, Scott and Julie Roederer, Scott Richardson, Anita Runge, Bill Schmoker, Jim Smith, Andrew Spencer, Jeff Stevenson and the Denver Museum of Natural History, Carol Sullivan, Judy Visty, David Waltman, Susan Ward, Scott Weidensaul, Cole Wild, Judy Wright, The National Park Service at Rocky Mountain National Park (RMNP), and the individuals that are part of the list serve saw-whet.net

Part Two:

The Flammulated Owl

Part Three:

The Northern Saw-whet Owl

Part Four:

The Boreal Owl

Introduction

Small Mountain Owls

There are 19 species of owls in North America, 10 of which are considered either medium or large owls, with the largest being more than two feet from head to tail. The rest are considered small, with the smallest being the size of a bluebird.

Of the small owl species, I find four of them particularly interesting. For instance, imagine a nocturnal owl so tiny that it can comfortably perch on a number two pencil, or another small owl that's active during the day and has eyes in the back of its head, or a third species whose whistle-like toots often continue for an hour or more throughout the spring, and a fourth whose call sounds remarkably like that of a Wilson's Snipe.

These tiny predators are the Flammulated Owl, Northern Pygmy-Owl, Northern Saw-whet Owl, and the Boreal Owl, all of which have a similar range within the mountains of western North America.

Including the Boreal Owl, there are four species of small owls throughout most of the western mountains. The Northern Pygmy-Owl, Flammulated Owl, and Northern Saw-whet Owls are often found within the same habitat and have even been known to nest in the same tree simultaneously, and at times, even nest in different cavities within the same cavity, in different years of course.

The Northern Saw-whet Owls and Boreal Owls can, at times, nest in relatively close proximity; however, it is far less common to find Flammulated or Northern Pygmy-Owls nesting close to Boreal Owls.

After moving to Colorado in the late 1980s, I began assisting a researcher who studied Boreal Owls and other species native to Colorado.

Finding owl research fun and exciting, I took it upon myself to begin researching the Northern Pygmy-Owl, because at that time (the early 1990s) there was very little published information about the species. Knowing that Northern Pygmy-Owls nest throughout the mountains were I live, I found it a "no brainer" to begin studying them and, hopefully, add some insight into the bird's natural history.

Fledgling Flammulated Owl in Rocky Mountain National Park.

Northern Pygmy-Owl

My interest in the Northern Pygmy-Owl began even before I had ever seen a live bird. Having read a great deal about owls most of my life, I had some familiarity with this bluebird-sized predator. However, until I began watching them in the wild, in 1994, I had no idea that such a small bird could be so impressive.

This seven-inch owl is extraordinary for a number of reasons. It is anatomically atypical among most of the North American owls. Its ears are symmetrical like ours, the species flight is not at all silent, as with most owls, and they routinely capture and kill birds and animals larger and heavier than themselves.

I was so fascinated with the bird's small size and pugnacious personality that I immediately found them to be one of the most interesting creatures that I ever encountered.

Before my research began, I read everything I could find pertaining to Northern Pygmy-Owls, which, at the time, was not very much. It became quite evident to me that there were few researchers working with this species. As I began watching these owls in the wild, I became captivated by them.

Since 1998, I've found Northern Pygmy-Owls nesting in and around Rocky Mountain National Park (RMNP). Then in the winters of both 1999 and 2006 I had two different Northern Pygmy-Owls show up in my yard nearly every morning and afternoon, at which time I released live mice for the owls to capture and feed upon throughout the winter.

I also devised a trap that would guarantee the capture of the birds. This way I could band them in order to gain insight into the birds' daily and seasonal movements and hopefully document a longevity record for the species.

After finding my first nest in 1998, I wrote a paper entitled *The Northern Pygmy-Owl in Rocky Mountain National Park* that was published in the *Journal of the Colorado Field Ornithologists*.

As with the Flammulated and Northern Saw-whet Owls, I've had the opportunity to rehabilitate injured Northern Pygmy-Owls. These stories are near the end of each species' section.

As you read through the Northern Pygmy-Owl section of the book, you will, at times, read the words pygmy-owl(s). This pertains to both the Northern Pygmy-Owl and the Ferruginous Pygmy-Owl.

Flammulated Owl

Flammulated Owls are unmistakable little owls. Their dark eyes, ear tufts, and small size make them easy to identify. Like Northern Pygmy-Owls, Flammulated Owls are about bluebird size, but unlike the Northern Pygmy-Owls, Flammulated Owls are extremely nocturnal, becoming virtually inactive during the day.

Their vermiculated gray coloration enables them to remain concealed during the day while perched within dense vegetation or against the trunk of a conifer.

These owls arrive on their North American nesting grounds sometime in April or May. Their soft hoots can be heard on spring evenings as the males solicit females. Like most small owls, Flammulated Owls are highly ventriloquial, appearing to be calling in one place while actually being in a different spot entirely.

As with other small owls, Flammulated Owls are cavity nesters, preferring to nest in cavities excavated by Northern Flickers. As for their food preference, these owls are almost entirely insectivorous. Watching the adult birds delivering food to their nestlings is quite impressive. It is not uncommon for the male to deliver a moth or beetle to the awaiting female every minute for an hour or more. However, being insectivorous forces these owls to migrate south in the fall, wintering in Mexico and farther south.

In 2004, after finding the first nesting Flammulated Owl in RMNP, I again wrote a paper that was published in *The Journal of the Colorado Field Ornithologist*. This paper was entitled *Small Owls of Rocky Mountain National Park*.

Northern Saw-whet Owl

I was introduced to the Northern Saw-whet Owl in 1986, after assisting with a banding project in central Wisconsin. Unlike the Flammulated Owls and Northern Pygmy-Owls, Northern Saw-whet Owls are found throughout parts of east, central, and western North America.

These nocturnal owls seem to have a kind of Dr. Jekyll and Mr. Hyde personality, roosting quietly during the day, often in dense foliage, then after dark becoming incredible flying mouse traps, often catching several mice each evening.

These highly migratory owls move en masse south in the fall, with many being captured and banded by researchers, including myself, trying to gain insight into the birds' biology. Then in the spring, the owls make their annual pilgrimage north to their nesting grounds. However, some males tend to remain on the nesting grounds throughout the year.

When vocalizing, both the Northern Pygmy-Owls and Northern Saw-whet Owls sound remarkably similar, with both species having a single note "*toot*" as their primary advertising call. Within the eastern parts of their range, Northern Saw-whet Owls are easily identified, as they are the only bird making that type of call. However, in the west there are a number of birds that make a similar sound and, when vocalizing during the day, the Northern Saw-whet Owl can be a bit harder to identify. Fortunately, Northern Saw-whet Owls seldom call before dark.

Over the years I've had the opportunity to rehabilitate several Northern Saw-whet Owls. In fact, the very first injured bird that I had received, back in 1994, was a Northern Saw-whet Owl that was attacked by a house cat. Since then, several Northern Saw-whet Owls with various injuries have found their way to me, most of which, I'm happy to say, have been successfully returned to the wild.

I've received more injured Northern Saw-whet Owls than both Northern Pygmy-Owls and Flammulated Owls combined, which may be due in part to the Northern Saw-whet Owls being more numerous than Flammulated Owls. Migration of Northern Saw-whet Owls and Flammulated Owls appears to be more dangerous than the Northern Pygmy-Owls' sedentary preference.

The Boreal Owl

My first encounter with this medium-sized owl was in Northern Colorado in 1996. It was my introduction into owl research. I was assisting Dr. Ron Ryder, Professor Emeritus from Colorado State University, with his owl project, which he had begun in the early 1980s. These larger relatives of the Northern Saw-whet Owl are found within the boreal forests of the world.

Prior to 1963, the species was considered a winter visitor to the continental United States. But on 14 August 1963, a juvenile female was shot in Larimer County, northern Colorado. After intense study in the Rocky Mountain region, it was found that the species does in fact breed in the lower 48 states, with birds having been either seen or heard in Colorado, New Mexico, Wyoming, Montana, Idaho, and Washington state (Palmer, Ryder, 1984, O'Connell 1987).

These owls appear to be more migratory in the east than they are in the west with a number of the owls shorter than the others being trapped and banded each fall/winter in the mid west.

Due to my lack of first hand experience with Boreal Owls, this section of the book is a bit shorter than the others.

The Species

While researching Northern Pygmy-Owls, I began finding Northern Saw-whet Owls within the same areas that I was finding Northern Pygmy-Owls. Then later I read an article by Norton and Holt (1982) describing a Northern Pygmy-Owl and a Northern Saw-whet Owl nesting in the same tree at the same time.

In 1998, I found my first Northern Pygmy-Owls nesting in an aspen tree. The pair used that same cavity the following year. Then in 2004 a Flammulated Owl nested in that same cavity. In 2005, another Northern Pygmy-Owl nested within that tree in a different cavity. Furthermore, in 2007, a Northern Saw-whet Owl nested within that same tree in a cavity a little higher up. Northern Saw-whet Owls often nest higher than both Flammulated Owls and Northern Pygmy-Owls.

In 2006, I found a Northern Saw-whet Owl nesting just a few feet from a tree that a Northern Pygmy-Owl had nested in the previous year. This suggests to me that all three species prefer similar, if not the identical habitat for nesting. From 1996 through 2007, I found Northern Saw-whets and Boreal Owls calling within close proximity to one another, which tells me that those two species, at times, may nest in relatively close proximity.

At least in the west, these four species are, at times, found within the same habitat, yet for the most part prey on different creatures. They all feed on insects, but only the Flammulated Owl does so with exclusivity. The Northern Saw-whet Owls, Northern Pygmy-Owls, and Boreal Owls feed on mice, voles, and birds; however, the Northern Pygmy-Owls feed on birds more often than Northern Saw-whet-Owls and Boreal Owls do, and conversely, the Northern Saw-whet Owls and Boreal Owls feed more on mammals than Northern Pygmy-Owls do.

When searching through habitat that I feel is adequate for one or more of the species, I routinely find them (with the exception of the Boreal Owl) within a small section of aspen, Ponderosa Pine, and Douglas fir trees that have a water source, juniper bushes, and downed logs. This microhabitat has everything the birds need for nesting within a small pocket of their territory.

Part One:

The Northern Pygmy-Owl

Glaucidium gnoma

Chapter One

The Gnome of the Forest

Throughout North America there are 19 species of owls. The larger of these are more than two feet (310cm) from head to tail, and the smallest is about the size of a White-breasted Nuthatch. Owls inhabit virtually every habitat in North America, with the exception of the taller mountain peaks of the west.

Some species can be found nesting comfortably within city limits, while others prefer to inhabit areas far from any human habitation. There are species that favor the eastern United States and others that are found only in the west.

Two of the obligate western species are the Northern Pygmy-Owl and the Ferruginous Pygmy-Owl. Being approximately six-and-a-half (16.07 cm) to seven-and-a-half inches (19.05 cm) from head to tail makes these pygmy-owls two of the tiniest owl species in North America.

Pygmy-owls are so small in fact that the only North American owl smaller is the five-and-a-half inch (14.1cm) Elf Owl, which is found in southern Arizona, southwestern New Mexico, and south Texas.

In his book *Owls of the World*, author Dr. James Duncan suggests that there are possibly more than 30 species of pygmy-owls throughout the world, all of which are in the genus Glaucidium. The owl I'm concentrating on for this section of the book is the Northern Pygmy-Owl, whose scientific name is *Glaucidium gnoma* (which includes several races).

Glaucidium, from the Latin *glaukidion,* means a kind of owl with glaring eyes and *gnoma* from the Greek means a dwarf, goblin, or (in this case) a small bird of disproportionate features.

The majority of Northern Pygmy-Owl sightings appear to be in late fall and winter, when some birds, presumably young of the year, are hunting near homes, observing bird feeders, and searching for an unsuspecting songbird or vole that they can snatch.

Anatomically Atypical
Among Owls

Unlike most owls, pygmy-owls are equipped with comparatively short, rounded wings and very long tails. The wings are so short, in fact, that when perched, their primary flight feathers end just a few millimeters past their rump. The tail of a pygmy-owl is quite long, extending almost three inches past its rump.

This wing and tail ratio is much like that of the Sharp-shinned Hawk or other accipiter. Both the accipiters and pygmy-owls often hunt in densely wooded areas. Therefore, having this type of anatomical structure allows these diurnal predators to move rapidly through the woods, during the day, when attempting to capture fast-moving prey and/or evading a larger predator.

{As a side note, the nocturnal owls (most of which are migratory), i.e., Burrowing, Flammulated, Northern Saw-whet, etc. have large broad wings and short tails. They obviously do not need the rapid flight ability of the pygmy-owls after dark}.

The flight of the nocturnal owls is silent and virtually inaudible to the human ear, because the edges of their flight feathers are fringed like a comb. These comb-like edges are most evident when looking at the bird's outer primary flight feathers.

This adaptation enables air to flow through the edges of the feathers as opposed to around the edges as it does with magpie or grouse feathers, for example. Having fringe-like feathers muffles the owls' flight, making them virtually silent fliers.

Pygmy-owls, being crepuscular (active in early morning and late afternoon), have a flight that is not at all silent. Presumably there is no need for pygmy-owls to have a silent flight when flying during the day.

The flight of the Northern Pygmy-Owl is a direct undulating burst of rapid flapping and short glides reminiscent of that of a woodpecker or shrike. Incidentally, when they land, it's most often with a hard-hitting "thud."

An adult Northern Pygmy-Owl. Note the short wings and long tail © Steve Morello

Juvenile Sharp-Shinned Hawk. Both accipiters and the Northern Pygmy-Owls have short, rounded wings and a long tail.

Therefore, this anatomical structure, along with the bird's small size, makes pygmy-owls unmistakable. In fact, throughout western North America, the only owl that could be mistaken for a Northern Pygmy-Owl would be the Ferruginous Pygmy-Owl. And the only place in the lower 48 states where these two species are found in relative close proximity is southeastern Arizona. However, in Arizona, the northern is found at higher elevations than the ferruginous.

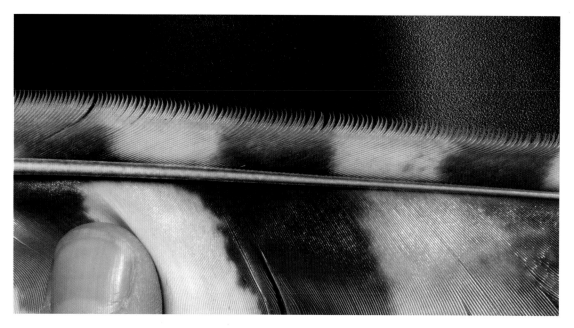

Close-up of the leading edge of a Great Horned Owl's outer primary flight feather. As the owl flies, the air flows through the edges of the feathers allowing the bird to fly virtually silent.

Close-up of the leading edges of the primary flight feathers of a Northern Pygmy-Owl. As the owl flies, the air flows around the edges of the feathers, making the bird's flight rather noisy for an owl.

Color Phases

The overall color of Northern Pygmy-Owls is either dark gray, rufous or brown, with the darkest brown populations found in Vancouver Island and British Columbia. The palest gray birds are found throughout the Rocky Mountains (Holt and Peterson 2000).

Their breasts and bellies are white with dark vertical streaks. Their flanks and backs are dark with small white spots. Their backs and wings are often covered with small white spots as well. The bird's long tails have several small horizontal bars and a white tip (when fresh). Their legs are feathered to the toes, but their yellowish toes themselves are virtually unfeathered and their talons (claws) are comparatively long and razor sharp.

Eyes, Ears, and Mouth

Like all owls, pygmy-owls have comparatively large heads, forward facing eyes, and binocular vision. Between the pygmies' piercing yellow eyes are large horned-colored bills. Surrounding the eyes are loosely woven facial disks, which act as a kind of sound receptor funneling sound to the birds' ears.

The ears of pygmy-owls, like all owls, are on the sides of their heads; however, unlike most other owls, the ears of pygmies are virtually symmetrical. And because they hunt during the day, detecting their prey by sight rather than by the sounds it emits, pygmies don't have or need asymmetrical ears, or tightly woven facial disks. Pygmy-owls may actually have the more poorly defined facial disks of all owls. This tells me that they locate their prey by visual cues rather then audial ones.

{As a side note, most owl species have well-defined facial disks and ears that are asymmetrically arranged. They have one ear that is larger and higher than the other. This enables sound waves emitted by their prey to reach the owls' ears at two slightly different intervals. The owl, then, can pinpoint its victim just by the sounds it makes.}

Interestingly enough, pygmy-owls may have the poorest night vision of all North American owls. But, their vision during the day is as acute as any hawk or falcon, in my opinion.

I have watched pygmy-owls, both northern and ferruginous, on numerous occasions, perch conspicuously on a bright sunny day. More often than not, in that case, the owls will perch in such a way as to use the shadow of an adjacent branch to shade their eyes.

As for owls' color vision, studies suggest that they see primarily in black and white, but may see a few colors, probably blues, yellows, and greens (Toops 1990).

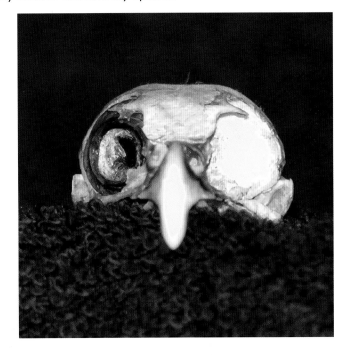

Note the symmetry of the ear openings. The owl's ear canals are just below and slightly behind the eye sockets. The left orbital ring is still attached to this skull.

Pygmy-Owls locate their prey by sight, more so than by the sounds it emits. Therefore, they have evolved with what may be the poorest defined facial disk of all North American Owl species.

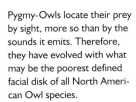

This owl was perched along a small meadow in Rocky Mountain National Park hunting small mammals. When they hunt mammalian prey, they are often perched in the open so it is easier for them to scan for prey.

Tufts, Flight Description, and Eye Spots

Other physical difference adding to the North American pygmy-owls'(both northern and ferruginous) uniqueness include their ear tufts (sometimes referred to as horns) and relatively noisy flight.

Ear tufts are sets of feathers on the top of an owl's head above its eyes that can be raised and lowered at will. When these tufts are raised, they help break the owls' silhouette. They also aid with camouflaging the birds during the day and can hopefully deter any potential predator from identifying the owl when it's roosting.

When Northern Pygmy-Owls are trying to hide, they often raise their ear tufts in an attempt to conceal themselves.

These ear tufts may actually make the face of an owl appear to mimic the face of a mammal such as a fox or bobcat. When a mammal confronts an eared owl, the owl raises its tufts and the predator may actually believe it is confronting another mammal and, not wanting the confrontation, may withdraw from attack (Mysterud and Dunker 1979).

Relatively speaking, owls either have ear tufts, such as the Great Horned Owl, Long-eared Owl, and Short-eared Owls, or they have no ear tufts, such as the Barn Owl, Northern Saw-whet Owl, and the Elf Owl.

Interestingly enough, pygmy-owls have evolved somewhere between these two groups. They have tiny feathers on the outer corners of their facial disk that can be raised when a bird is trying to conceal itself.

On several occasions, I've witnessed Northern Pygmy-Owls raising their little horns. Each time, the Northern Pygmy-Owls were perched on an exposed limb as a songbird landed nearby. The owl raised its horns and snapped its bill a few times as the bird moved off.

The Great Horned
Owl is an example of a
tufted owl.

Small Mountain Owls · Part One

As Northern Pygmy-Owls perch conspicuously, searching for an unsuspecting vole or other small mammal, they are frequently harassed by songbirds. Songbirds seemingly know that the owl can be, and often is, a potential threat to them. The harassing birds most often fly around the owl, scolding it angrily, but seldom if ever actually strike it. The reason the attacking birds don't actually strike the owl may be due in large part to the owls' false eyes on the back of its head. All pygmy-owls have feathered eye spots on the back of their heads that consist of two horizontal black, tear-drop shaped spots outlined in white. These eyespots are quite a bit larger than the owl's real eyes, and from a distance looks more like eyes than the bird's real eyes do.

These spots may actually deter predators from attacking the owl from behind because the owl appears to maintain a constant vigil as it rapidly swivels its head while monitoring an onslaught.

To me, these false eye spots may be the owls most interesting adaptation.

The Barn Owl is an example of a species without ear tufts.

Left: All Pygmy-Owls species have eyespots on the back of their heads. From a distance, these eyespots look more like eyes than the bird's real eyes do. *Right:* when the birds get excited they often twitch their tail from side to side in a quick jerking motion.

The Northern Pygmy-Owl   23

As the female Northern Pygmy-Owl enters her nest, note how abraded the tip of the tail is. While she is inside the nest, her tail scrapes against the walls of the nest cavity. When you notice this, you can discern the female owl from the male.

The Difference Between Males and Females

In the wild, it can be difficult to discern the male Northern Pygmy-Owl from his mate. However, the females are a bit larger than males, but to be honest, it is only noticeable if the pair is side by side. Occasionally, this size difference is very slight. However, males are on average grayer and females average more red and browner (Pyle 1997).

While talking to Kay McKeever of the Owl Foundation in Ontario, Canada, she explained to me that it's not the difference in the birds' size that is important to their relationship, but rather that the female Northern Pygmy-Owl is much more aggressive than the male.

For the most part, female owls are usually larger than males, a term known as reverse sexual dimorphism. The significance of this is still under debate with most ornithologists. However, some reasons for the size difference may be that having adults of two different sizes enables the smaller male to hunt smaller prey (specifically during the nesting season) which is often more abundant and easier for him to overpower. Then, as the little ones grow, the female can capture larger prey that will satisfy the owlets' increasing appetites. Furthermore, having adult birds of two different sizes enables the pair to hunt the same area without competing against each other for food (Storer, 1952-1966 & Selander, 1966).

Territory Size

Before I began studying Northern Pygmy-Owls, it was suggested to me that they may have the largest territory of any small owl. As I found during my study, this seven-inch (17.78 cm) predator does, in fact, have a rather large territory.

One pair had a territory that was about 190 acres. To me, this is quite impressive when you think that this seven-inch (17.78 cm) owl has a territory size similar to that of a 23 inch (58.42 cm) Great Horned Owl. Interestingly enough, there were several occasions during the study when both species were found within relatively close proximity to one another.

When determining the size of the Northern Pygmy-Owls' territory, I made mental notes of where the birds were seen during the courtship and nesting. Then during the fall, after I had lost track of the birds, I walked what I believed to be the owl's territory with a GPS (Global Positioning System). After I had walked the territory, I took the GPS back to the national park headquarters and entered the information into their computer-mapping program and the computer calculated the size of the bird's territory, including elevation gain. Ironically, during that hike, I came across a Northern Pygmy-Owl perched on a limb at the northern edge of its territory.

In Washington state, non-nesting Northern Pygmy-Owls, including owls that attempted nesting and failed, were found ranging from a few kilometers to more than 15.5 miles (25 km) for a few days to as many as 30 days before returning to their nesting area (Giese, Arnheim & Forsman 1997).

However, during the summer months, these owls seemed to keep a relatively contiguous home range. One owl (prior to nesting) made an excursion of more than six miles (10 km), staying away for two weeks before returning to its eventual nest area prior to the start of incubation. The bird returned between 11 and 19 days before incubation commenced (Giese, Arnheim & Forsman 1997).

Throughout the owls' North American range, they have been found from sea level on the West Coast to as high as 12,000 feet (3659 m) in the Rockies.

Distribution and Range

According to the Bird Banding Laboratory and the A.O.U. (American Ornithologist's Union) check list of the birds of North America there are two species of pygmy-owls found in North America: the Northern Pygmy-Owl and the Ferruginous Pygmy-Owl.

The Northern is found from southeastern Alaska, down through British Columbia, Southwestern Alberta, south through the foothills and mountainous regions of the West to Southern California, and east to Central Colorado. It is also found from central New Mexico and extreme west Texas down through central Mexico to Guatemala (Johnsgard, 2002).

However, throughout their range, Northern Pygmy-Owls are found from approximately 6000 to 12,000 feet (1968 m to 3659 m) above sea level. Being nonmigratory, some individuals can often be found within their territories year round.

The Ferruginous Pygmy-Owl, on the other hand, is found in south-central Arizona and southern Texas, south to at least northern Argentina (Johnsgard 2002).

Fossil records of pygmy-owls in the Americas are found as far back as the Pleistocene epoch, which was 1.8 million years ago (Brodkorb, 1971).

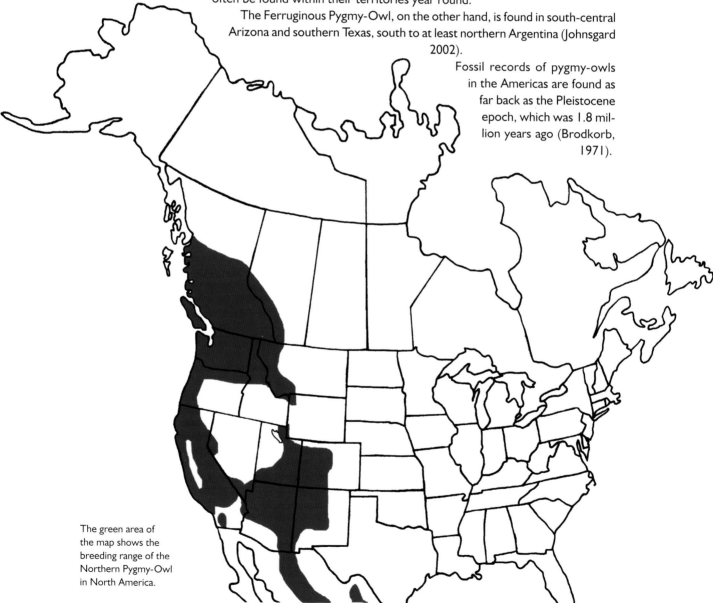

The green area of the map shows the breeding range of the Northern Pygmy-Owl in North America.

Chapter Two

Vocalization

{The territorial calls of the larger North American Owls, i.e., Great Horned, Barred, and Great Gray, consist of deep resonating hoots that can carry over a mile on a windless evening. These hoots can be heard at any time of the year but are most often heard from late fall through spring when the individual owls are advertising their presence for a potential mate or to bond with their existing one. If you've spent any time outdoors from late winter to early summer, you've probably heard one or more of these larger species calling.

In contrast, the majority of the smaller owls (with the exception of the Flammulated Owl) don't hoot at all. In fact, their calls are anything but hoots. One call of the Eastern Screech Owl, for example, sounds (to me at least) very much like the whinny of a small horse and the Burrowing Owl even mimics the rattle of a rattlesnake when closely approached.}

Most often, when a Northern Pygmy-Owl vocalizes, it will lean forward and toot. With each toot, the bird pumps its tail.

The territorial call of the Rocky Mountain race of the Northern Pygmy-Owl is most often a hollow, whistled, single note *toot* which is often given in a rhythmical succession reminiscent of *toot------toot------toot------toot------toot,* etc. Each toot is often spaced one to two seconds apart. The bird has been known to carry on for up to five minutes at a time and sometimes even longer. Burroughs (1906) wrote about the call of the Northern Pygmy-Owl: It was such a sound as a boy might make by blowing in the neck of an empty bottle and a Mrs. Bailey (1928) wrote that the sound of an immature female pygmy sounded like a long whistle followed by a cuckoo-like cuck-cuck-cuck-cuck.

The territorial call of the Northern Pygmy-Owls on the West Coast is slower in tempo than the birds of the inner West. Birds in the northern portions of the owls' range often have a single note call, where the birds in the southern part of the range (southeast Arizona for example) have a double-note call (Howell and Webb 1995, Holt and Petersen 2000). I have heard the owls in Colorado give both single and double note calls.

To some researchers, this slight vocal variation along with some subtle plumage differences is enough to separate the Northern Pygmy-Owl into three species. If that is all it takes to separate species, then I would have to make the argument that throughout North America, the Great Horned Owl should be separated into different species for the same reasons. However, there are different races of Great Horned Owls, but no matter which Great Horned Owl you see, or hear, they're all the same species, *Bubo virginianus*.

On a windless day, the call of a Northern Pygmy-Owl can carry roughly a mile. When vocalizing, Northern Pygmy-Owls frequently perch on an exposed limb (a dead branch) anywhere from a few feet off the ground to as high as 60 feet (18.29 meters) or more. The birds face forward, raise their heads, and *toot*, revealing their attractive white necklaces. With each toot the birds pump their tails. While calling, especially on an exposed limb, Northern Pygmy-Owls remain constantly vigilant, surveying their surroundings for potential danger or a potential meal.

Their call at times has a ventriloquial quality as well, making them somewhat difficult to pinpoint. Besides the Northern Pygmy-Owl, I have heard this ventriloquial quality in calling Boreal Owls , Flammulated Owls, and Northern Saw-whet Owls as well.

Northern Pygmy-Owls can be heard calling any time of the year, with their calling becoming most persistent from late winter (February) through the beginning of incubation (April through June).

Marking His Territory

While researching Northern Pygmy-Owls in Rocky Mountain National Park (RMNP) one late afternoon (in 1999), I heard the male calling at the western edge of his territory about 40 minutes before dark. He had been vocalizing for quite a while before I found him. A few minutes after the bird was located, it stopped vocalizing, then flew about 50 yards (45.50 m) east and began calling again.

This calling bout had lasted longer than I had previously heard, so I decided to find out how long it lasted. Each time the bird called, it continued almost five minutes, and then stopped. Then he moved roughly 50 yards (45.50 m) to the east and repeated the same routine. After the third consecutive calling bout, the owl flew across the meadow to the north side of the trail and continued this activity three more times moving east each time before he stopped calling. By this time it was quite dark and I believe the bird stopped vocalizing due to the darkness.

Northern Pygmy-Owls often begin calling before sunrise and continue through mid-morning, then seem to take a break during midday and resume calling in the early evening, continuing until dark. One full moon evening in April, I heard a Northern Pygmy-Owl calling on and off until 11:00 p.m., which to me seemed a bit unusual, but I suspect that the light of the moon had something to do with this. Most often, the owl will cease calling just before it is too dark to be seen. They are also more apt to be calling throughout the day if that particular day is overcast rather than sunny.

Other Birds that Sound Like the Northern Pygmy-Owl

During the day, the Northern Pygmy-Owls' calls (whistled *toots*) can be, and often are, mistaken for the similar single call notes of the Townsend's Solitaire but, the call note of the Solitaire is higher pitched and more sporadic than that of the owl. (The Townsend's Solitaire is a beautiful medium-sized gray thrush, slightly larger than a bluebird, and quite plentiful in the same habitat as the Northern Pygmy-Owls, and both species are found throughout the Rocky Mountains year round).

Interestingly enough, Common Ravens and American Crows calling off in the distance can, on occasion, sound remarkably like the calling of the Northern Pygmy-Owl.

After dark, a calling Northern Saw-whet Owl can be mistaken for a Northern Pygmy-Owl too. The tonal qualities of these two species' voices are very similar, but there are a few distinct differences between them.

Townsend's Solitaires are a member of the thrush family. Their call sounds similar to the call of the Northern Pygmy-Owl.
© Bill Schmoker

The most obvious difference is that Northern Pygmy-Owls are often calling well before sunset, and end just after it is totally dark; whereas the Northern Saw-whet Owls usually don't start calling until just before dark, often continuing several hours through the night, and may even continue until just before sunrise.

Furthermore, when Northern Saw-whet Owls call, they often remain vocalizing for up to and longer than an hour while Northern Pygmy-Owls seldom vocalize for more than about five minutes before stopping.

In April 2004, I came across a calling Northern Saw-whet Owl only 100 feet (32.80 m) from an abandoned Northern Pygmy-Owl nest, but that nest was used by a pair of Flammulated Owls that year (which I will detail in Part 2).

Incidentally, Bailey and Neidrach (1962) and myself, in 2004, found Northern Pygmy-Owls breeding in the same areas as Northern Saw-whets and Flammulated Owls.

One late afternoon, just after the sun had set, I located the male Northern Pygmy-Owl perched in the top of an aspen giving his distinctive toots. While watching and listened to the little owl, off in the distance I heard the *who.. who..who whooo..* of a Great Horned Owl. The little Northern Pygmy-Owl stopped calling, flew to a higher perch, faced the direction of the larger predator, and started vocalizing, as if to say, "this is my territory, stay out if you know what's good for you."

On two occasions (2005), I heard a male Northern Pygmy-Owl calling just a few hundred yards from a calling Great Horned Owl. Fortunately, there was no interaction between the species because the larger owl would have no trouble dispatching the smaller one.

As with the solitaire, the Northern Saw-whet Owl has a call that sounds similar to that of a Northern Pygmy-Owl.

Variations in the Northern Pygmy-Owl's Call

The voice of the Northern Pygmy-Owl can vary widely depending on the situation. For example, the male's single note calling often becomes coupled when he gets excited. On other occasions, calling birds will vocalize with a short burst of rapid *toots* followed by a series of single ones. It's common to hear the birds calling in a somewhat sporadic nature as well, but this is a bit less common.

A circumstance when the male might call sporadically would be, for example, when the female comes close to him the first few times during courtship, and when the female enters the nest cavity for the first few times before incubation begins.

As nesting season approaches, both male and female Northern Pygmy-Owls become quite vocal. As I am attempting to locate their nest, I listen for and follow the female because she will eventually lead me to it.

When discerning male from female Northern Pygmy-Owls (by voice), it's handy to know that the female's voice is higher in pitch than the males.

By May, any Northern Pygmy-Owl that is still calling is most likely an unmated individual. Northern Pygmy-Owls will, however, start responding to imitation of its calls after about mid-August.

In contrast to the territorial calls, the voice of an agitated Northern Pygmy-Owl is a kind of strange series of "eyrt" notes, which to me is reminiscent of a fingernail moving across a chalkboard.

Throughout my research, I have found several interesting aspects of the Northern Pygmy-Owl's personal lives. A few of these came to light while monitoring the owls, before the female began incubating her eggs.

One season, while watching and listening to a male Northern Pygmy-Owl calling, his mate was seemingly responding from about 100 yards, (91 meters) off. Then a third owl began calling. As I watched and listened to the first owl, it stopped calling momentarily, presumably listening, (possibly) for its mate and also to figure out how close the intruder might be. After a few moments the original owl continued calling. I presumed the territorial male did not consider that third bird to be enough of a problem to warrant a confrontation. It was suggested to me that the third bird might have been one of the pair's offspring from a previous nesting season, or possibly an unmated bird from an adjacent territory.

In 2005, I witnessed the pair duetting. While researching the birds that spring, I made my way towards a calling bird. As I moved closer to it, a second bird began vocalizing in the same area. I approached the birds, as their voices melded together and became perfectly stereophonic. Until I heard this, I was unaware that the birds vocalized in this manner. After a minute of so, one bird flew off as the other followed.

Once the female accepts the nest site, she spends most of her time in relative close proximity to it for several days before incubation begins. Once the eggs were laid and incubation was underway, the Northern Pygmy-Owl pair seldom, if ever gave their territorial calls or responded to imitations of it.

As an interesting sidebar, temperature and weather seem to be of no concern to Northern Pygmy-Owls. I have heard them calling in a variety of weather conditions, from heavy snowfall to light and even moderate rain as well as temperatures well below zero.

Courtship

Due to the Northern Pygmy-Owls' small size and rather large territory, there is very little documented on the birds' courtship activities. However, it has been suggested that Northern Pygmy-Owls' courtship can last as long as eight weeks (Cramp 1985).

While in RMNP in the spring of 2005, I witnessed a pair of Northern Pygmy-Owls courting and the description is as follows. At 6:07 p.m. two owls are calling approximately 400 yards (364 meters) apart. The first owl, presumably the female, was perched in a dead aspen roughly 25 feet (7.58 m) off the ground. It was giving some soft cooing calls and soft twittering. I identified this owl as the female because her voice is slightly higher than his. Interestingly enough, while the owl twitters, her entire body quivers (this cooing is reminiscent of a calling Mourning Dove and the twittering sounds like that of a Chipping Sparrow's territorial call).

At 6:15 p.m., the male came within view of the female with a vole in his right foot. He waited three minutes perching approximately 12 feet (3.64 meters) from her. At 6:18 p.m. the male flew within a few inches of the female, she cooed and looked down. The male moved behind and copulated with her. During this act she twittered then cooed, the male gave her the vole, and he flew off.

At 6:21 p.m. the male was an estimated 50 yards (45.50 meters) from his mate, calling.

Food transfers and copulations were witnessed a couple of times before nesting started. On a few occasions the copulation occurred without a food transfer. This similar account was witnessed by Mr. and Mrs. C.W. Michael (Bent 1938). In that account, a long trilling note was heard given by one bird, and soon, a second bird had arrived. Both birds were seen snuggling together. The male gave a vireo to the female.

The courtship of the California race of the Northern Pygmy-Owl was witnessed by a John Lord in 1866. He writes: "In the evening twilight the owls again come out of their hole and take erratic flights around their abode, chasing each other up and down the plain, and performing all kinds of inexplicable maneuvers. Occasionally they settle on the ground, but never long at a time".

On the morning of 20 April 1929, a Mr. C.W. Michael watched a pygmy-owl enter a nest cavity and soon after, heard a second owl begin a soft fluttering trill. After locating the second owl, roughly 10 feet (3.05 m) from the nest, the female peered from the nest hole, waited for a few moments, then exiting the nest and flew to the calling bird. After a few moments the birds accomplished the supreme embrace (Bent 1938).

Copulation of Northern Pygmy-Owls.

Female Northern Pygmy-Owl perched near her nest, giving her food begging call. Note the frayed tip of her tail. © Jim Osterberg

Food Begging Calls of the Adult Female

The term food begging refers to the unique sounds that a creature makes when it's hungry and wants its parents or mate to feed it.

After nesting has begun, the food begging calls of the female Northern Pygmy-Owl consist of a short sequence of high pitched twittering notes.

When hungry, the female will often beg with a series of five-to-seven rapid *kewing* notes followed immediately by three to five slower ones, such as..... *kew,kew,kew,kew,kew-------------kew-----kew-----kew-----kew.*

Food begging is most often heard in a series of three to six rapid notes followed by three or four well-spaced ones and is most often heard several days prior to incubation. This calling apparently means that she has agreed to the nest her mate has chosen. This particular food begging call is apparently given by the female just before and just after the young hatch.

When the male arrives near the nest with food, he will utter a series of *toots*. Usually one-to-three notes, which apparently tells his mate that he has arrived with food. She then, more often than not, comes to him to retrieve what he has brought for her.

Danger Calls

When the male believes danger is approaching either his mate or the nest tree, he will utter five rapid toots. I witnessed this on three different occasions.

The first occurred one mid-afternoon. While both adult owls were perched near the nest (in different trees) the male gave a rapid five-note-toot and, instantly, the female flew straight into the nest. I didn't see what made him make this call, though.

On another occasion, both male and female were perched near their nest and a Peregrine Falcon flew overhead. The male tooted twice as if to say look out and once the falcon passed, the female tooted once as if to say I saw it. This same calling sequence was witnessed the following year when a Cooper's Hawk flew over the nest tree.

Chapter Three

Monitoring Nesting

Finding the territory of a Northern Pygmy-Owl is not as difficult as one might think. Beginning around mid-February, the males begin vocalizing with the intention of attracting females. As the males announce their presence, they move throughout their domains, seemingly leaving the vocal equivalent of scent marks throughout their territories.

Within Colorado, the preferred habitat of the Northern Pygmy-Owl appears to be areas that include a mixture of pine, spruce, fur, and aspen along with a meadow and an active water source such as a creek or pond.

Being a secondary cavity nesting owl, Northern Pygmy-Owls need a number of tree cavities from which the males can choose an appropriate home.

The Northern Pygmy-Owl's nest tree is the second aspen from the right. The nest cavity was the hole in the center of the tree.

Something else that is helpful in locating these owls is the presence of songbirds that become agitated when the Northern Pygmy Owl's territorial call is broadcast.

Northern Pygmy-Owls are seemingly always being bothered by songbirds. As I was watching a calling owl one afternoon, I witnessed an interesting interaction between the owl and a mixed flock of Mountain Chickadees, Pygmy Nuthatches, and Ruby-crowned Kinglets. The owl was calling while perched within a live Ponderosa Pine. As I approached the calling owl, the much smaller songbirds vocalizing in protest already surrounded it.

The small birds flew around the owl vigorously for a few moments while the owl was vocalizing, but the owl continued for several minutes, paying the songbirds no mind. The flock slowly moved off and the little owl continued calling, obviously more interested in vocalizing than catching something to snack on. I have witnessed this same routine countless times over the years.

Searching for areas that include a combination of tree types, ground cover, and a water source will greatly increase your odds of locating one of these little owls.

Those first few years of my research, I would begin surveying on 15 February and would enter the woods around 3:00 p.m. because that gave me about two-and-a-half to three hours of research time before it becomes too dark to see the birds. Then of course as daylight lengthens, I began my investigation later each day because the owls tended to vocalize just before it was too dark to see them.

Where Do Northern Pygmy-Owls Nest?

Northern Pygmy-Owls, like the other small owls, are secondary cavity nesters. This means that they nest in a cavity, yet lack the knowledge or skill to excavate one themselves. Therefore, in order for Northern Pygmy-Owls to nest, they need a cavity excavated by a woodpecker or a man-made nest box in order to set up house keeping.

Incidentally, woodpeckers would be an example of primary cavity nesters because they excavate their own cavities.

Over the years, I've found Northern Pygmy-Owls nesting in cavities within both aspen and poplar trees. However, they have been documented nesting in other tree types, including oak, pine, and Douglas fir. In every case, the cavities were excavated by Northern Flickers, Hairy Woodpeckers, both Williamson's and Red-napped Sapsuckers, and/or Downy Woodpeckers.

Unlike the Northern Saw-whet and screech owls, I haven't found any documentation to support wild Northern Pygmy-Owls using nest boxes. However, Ms. Kay McKeever, president of the Owl Foundation, does have Northern Pygmy-Owls nesting in nest boxes in her facility in Ontario, Canada.

Like other North American owl species, Northern Pygmy-Owls haven't been documented adding or removing any material to or from their nests. However, on one occasion, I watched a female Northern Pygmy-Owl enter her nest with an unidentified songbird and, after a few moments, she was at the nest entrance spitting out a small mouthful of feathers. These feathers may have just been stuck to her bill when she came to the entrance and she merely spat them out.

On the other hand, the European Pygmy-Owl, a close relative of the Northern and Ferruginous Pygmy-Owls, will routinely sanitize its nest cavity. She (the female owl) "sanitates" (sic) the nest hole by carrying out food remains, which are dropped at the foot of the nest or elsewhere within the sojourning sector. At the very end of the nesting period, no sanitation takes place. (Erkki Kellomaki 1977). Neither the Northern nor Ferruginous Pygmy-Owls have been documented cleaning out their nests in this manner.

Another Northern Pygmy-Owl nest tree. The birds often nest at the edge of a wooded area, which enables them the luxury of hunting in a small open area, or in the thicker woods if needed.

The Northern Flicker is most often the architect of the nest cavities that the Northern Pygmy-Owl and other small owls use for nesting.

Nest Descriptions

Between 1998 and 2005, I have located Northern Pygmy-Owls nesting in the Cow Creek area of RMNP. The first two years (1998 and 1999), the pair occupied the same cavity during both nesting seasons.

That cavity (excavated by a Northern Flicker) was in a living aspen. The nest was 16 feet (4.87 meters) from the ground and had an entrance hole approximately two inches by two inches in diameter (5.08 by 5.08 cm). The west facing entrance was directed in such a way that allowed the female owl a direct flight to and from her domicile.

The nest tree was roughly 124 feet (37.61 meters) from a large meadow and 300 feet (91.46 meters) from a moving stream. Just to the south of the nest, extending up the ridge are several acres of beetle-killed spruce. These trees have countless limbs that the owls perched on while surveying their territory.

Female Northern Pygmy-Owl looking from her nest prior to exiting.

The center aspen was the nest tree that Northern Pygmy-Owls used for nesting in 1998, 1999, and 2004. Unfortunately, during the winter of 2005, the tree had fallen down.

The area also has several small shrubs and downed logs, making it an ideal area for mice, voles, shrews, and chipmunks, which the owls fed upon during those years.

I believe the same pair to be the individuals that nested in that cavity in 1998 and 1999, but I was unable to mark the adults to verify it. However, there were things that the female did both years that were identical. For example, she allowed me close proximity to the nest during the nesting season, and she perched on the same limb of a dead spruce while keeping watch over her nest.

On the other hand, there were a few things that had changed dramatically that second season. The most obvious was that the owls had a bunch of new neighbors that frequently harassed the adult owls every chance they got. These neighbors were Dark-eyed Juncos, Chipping Sparrows, Mountain Chickadees, House Wrens, American Robins, and MacGillivray's Warblers.

On several occasions, as the female Northern Pygmy-Owl exited her nest, she would instantly be badgered by these birds. It was obvious to me that neither owl nor songbirds knew the others were going to set up housekeeping in the area. Because the owls begin nesting before the songbirds do, I expect that the songbirds would not have chosen that area for nesting had they known the owl was already there.

Songbirds see the owl as a potential threat to both themselves and their nestlings and therefore the little birds disapproved of the owl wholeheartedly. By attacking the owl, the smaller birds are hoping to drive the larger predator from the area. Any songbird that finds an owl in the daytime will harass it, sometimes for several minutes, before moving away. The list of songbirds documented harassing Northern Pygmy-Owls reads like a field guide to songbirds of western North America.

The third nesting season (2000), the owls chose to nest in an aspen as well. Like the previous site, this nest was in an abandoned Northern Flicker cavity, but this one faced south. It was 14 feet, five-and-a-half inches (4.27 meters, 13.97 centimeters) from the ground and the entrance hole was two-and-a-half inches high by two-and-three-eighths inches wide (6.5 cm x 6.1 cm.).

Adult Northern Pygmy-Owl perched near her nest.

The nest tree is only 65 feet (19.72 m) from the backdoor of a cabin that, at the time, was being remodeled. It is also 90 feet (27.44 m) from a hiking trail that was in use daily. Oddly enough, as far as I could tell, no one but me knew the birds nested there. Even the construction workers had no idea the owls were present.

Due to the construction, there were few songbirds in the area, enabling the owls to go about their daily activities without harassment. The bird activity near the previous nest site may have upset the adults so much that they chose to nest in an area in which the adults and young had a very small chance of harassment by songbirds.

In my opinion, Northern Pygmy-Owls consider other birds and animals more of a threat to their young than human activity. When the female feels her young are in danger, she will not hesitate to attack any potential predator, including any bird, animal, or even people.

In 2005, I was able to locate two nests and, like the previous nests, those were in aspen trees as well. One was excavated by a Northern Flicker and was 24 feet (7.28 m) from the ground. As with the other nests, that nest had an entrance about two inches (5.08 cm) in diameter. Interestingly enough, that cavity was in the same tree that a pair of Northern Pygmy-Owls used during the 1998 and 1999 nesting seasons. In 2005, that nest was 25.5 feet (7.47 m) from the ground and faced north. It was the highest Northern Pygmy-Owl nest that I had found. It was about six feet (1.83 m) from a live spruce, which the adult owls used to conceal themselves before entering and after exiting their home.

The second nest found that year was excavated by a Hairy Woodpecker. It was quite a bit different than any previous nests that I had located. It was the lowest Northern Pygmy-Owl nest that I had come across up to that point and it also had the smallest entrance hole I had come across. It was only 11.66 feet (3.55 m) from the ground and had an entrance that was only one-and three-fourth inches (44cm) by one-and-seven-eighth inches (48cm) and the cavity itself was only seven-and seven-eighths inches (200 cm) deep.

Northern Pygmy-Owls, will on occasion, nest in close proximity to other birds. The above-mentioned nest was in the center of several nesting woodpeckers. The owls chose to nest in a cavity 63.5 feet (19.36 m) from an active Northern Flicker nest, 22 feet (6.71m) from an active Red Squirrel nest, and 111 feet (33.84 m) from an active Hairy Woodpecker nest; yet I saw no interaction between the species. In this case, I believe the owls had such limited nesting options that they took a chance to nest near these other birds and it worked out for everyone.

Every nest that I have located has had an active water source within a few yards of it. Therefore, a water source, i.e., creek, stream, river or lake, may be a deciding factor in where Northern Pygmy-Owls nest.

While in southeastern Arizona in the spring of 2005, I found a Northern Pygmy-Owl nest along a trail in Madera Canyon. Those birds chose to nest in a cavity within a sycamore tree. The tree and subsequent nest cavity were within a few feet of a hiking trail and as with the other nests that one was just a short distance from an active stream.

Throughout their North America range, Northern Pygmy-Owls have been found nesting from sea level in California (Bent 1938) to 490 meters (1600 feet) in British Columbia to as high as 3600 meters (more than 11,000 feet) in the central and northern Rocky Mountains.

Cleaning the Nest

Knowing that Northern Pygmy-Owls are not very tidy housekeepers, I decided to vacuum out the nest cavity after that first nesting season (1998). My intent was to identify any prey items that may have been inside after the young had fledged.

To do this I used a hand-held battery-operated vacuum with an extension tube which enabled me to reach the bottom of the nest to retrieve any prey remains still inside. These items are documented in Chapter Five.

Watching the Nests In and Around
Rocky Mountain National Park

As of the last week in February, the handsome Mountain Bluebirds begin arriving in Colorado from their wintering grounds, which could be anywhere from southern California to west Texas, or even Mexico.

Usually just a few brilliant blue males dot the landscape, then within the next few weeks more males are seen and the first females begin arriving. The bluebirds, especially the males, are welcome splashes of color in an otherwise drab landscape.

Also, during this time, the Northern Pygmy-Owls become much more vocal. And by late afternoon I'm most often able to find at least one male, and sometimes more, just by following their whistle-like toots.

Then by mid-March they are most reliably located by their vocalizing, which is often heard within close proximity to what will eventually be their nests. In those years when the owls nested in the same tree at Cow Creek, the adult birds could frequently be heard vocalizing southwest of the ranch across the meadow, up the hill within the woods.

The tree species in this part of the park are primarily Ponderosa Pine, aspen, spruce, and Douglas fir, along with a significant number of junipers with several downed logs interspersed. This area of the park has a stream that flows west to east and during the spring runoff can be quite deep and fast flowing.

The region where the owls were most often found, at least those first two years (1998 & 1999) and again in 2004, is most easily accessed by traversing a small section of the meadow, before the trail head, then up a slight incline before entering the woods.

The Cow Creek area of Rocky Mountain National Park. The creek itself flows through the willows. When the owls nested in the area, they could, at times, be seen perched along the willows or up on top of the spruce trees searching for voles.

Historically this portion of the national park was a working dude ranch, and there are still some remnants of the old trails that the wranglers used when the ranch was in its heyday. As you move through the woods, you will find one of these trails now riddled with downed logs and overgrown grasses. The only things using those trails now are the elk, deer, bears, bobcats, and Mountain Lions.

In early summer this trail is a great place within the park to readily find Calypso Orchids. These beautiful pink flowers grow in small bunches along the overgrown trail under the shade of the spruce trees. The orchids exist in this area because the elk are not found here during the flower's growing season. The elk find these flowers quite a delicacy and will consume any that they come across.

As I head west through the woods, I frequently come across small birds foraging in the trees along the trail. This time of year, the flocks often contain Ruby and Golden-crowned Kinglets, Brown Creepers, Mountain and Black-capped Chickadees, white-breasted, pygmy, and Red-breasted Nuthatches, and Dark-eyed Juncos. As the spring and the following summer progresses the bird list increases considerably as the migrants arrive.

Continuing west a few more yards, the trail meets the meadow and due to lack of use, the trail has reverted to grass. This tall grass reduces the chance of birders or hikers happening on that nest by accident. From that area, I would traverse a short distance through the mixed spruce fir forest before reaching the general vicinity where I would hear the owls.

The female Northern Pygmy-Owl could routinely be found within an area of roughly 90 square yards (81.9 m) for about two weeks before incubation. Within that general vicinity there are Ponderosa Pine trees and quite a few aspens complete with nest cavities excavated by the various woodpeckers of the area including red-napped and Williamson's Sapsuckers, Northern Flickers, as well as downy, hairy, and Three-toed Woodpeckers. This variety of nest cavities made it a bit tricky to initially locate which one the Northern Pygmy-Owls would eventually decide upon. But with persistence, and luck, I eventually found the nest.

Many areas of the forest floor are covered with cones from the various conifers in the area.

Mountain Chickadees are just one species that are found in the same areas as Northern Pygmy-Owls.

Calypso Orchids can be seen along the trail in the areas where the Northern Pygmy-Owls nest.

Embracing the Relationship

Studying Northern Pygmy-Owls, I've been fortunate to witness several food transfers between the male and female. When these transfers take place, the male moves near the female, placing the quarry in his bill, and uttering a few soft-coupled toots, which continues until the female comes to him.

Early one afternoon, I was present when the male Northern Pygmy-Owl arrived near the nest with a headless chipmunk, which he captured and beheaded before delivering it. He tooted softly as the female exited the nest and perched momentarily until she saw him. She flew to him moving slightly underneath. Reaching her legs out, she grasped the small mammal just as he dropped it, and then went directly into the nest to feed it to her nestlings.

Another transfer witnessed was with an unknown warbler. The female was once again inside the nest as her mate arrived with the small bird. He called with a series of soft toots, as the female came to him.

They perched side by side for a few moments, before he turned to her with the bird in his bill as she took it from him. She will often eat her fill then store the rest.

On a few occasions, I happened upon the Northern Pygmy-Owls just prior to consummating their relationship. The first time was 14 March 1998, at 6:03 p.m., I noticed an owl perching conspicuously on the limb of a dead aspen about 25 feet (7.62 m) from the ground. After a few minutes it began vocalizing while a second owl began calling from an estimated 100 yards (91 m) west. I identified the bird closest to me as the female by voice. As I mentioned earlier, the female's voice is higher pitched than the males.

Besides tooting, she gave cooing notes, and twitters as the male approached. At 6:15 p.m., I was within 25 feet (7.62 m) of her and she paid no attention to me. Then at 6:18 she cooed and leaned forward as the male came low behind her and they mated, she cooed and twittered as the male flew off. For the next few weeks this intimate embrace was observed almost daily until incubation began.

During courtship, the male Northern Pygmy-Owl brings back both birds and animals to his mate.

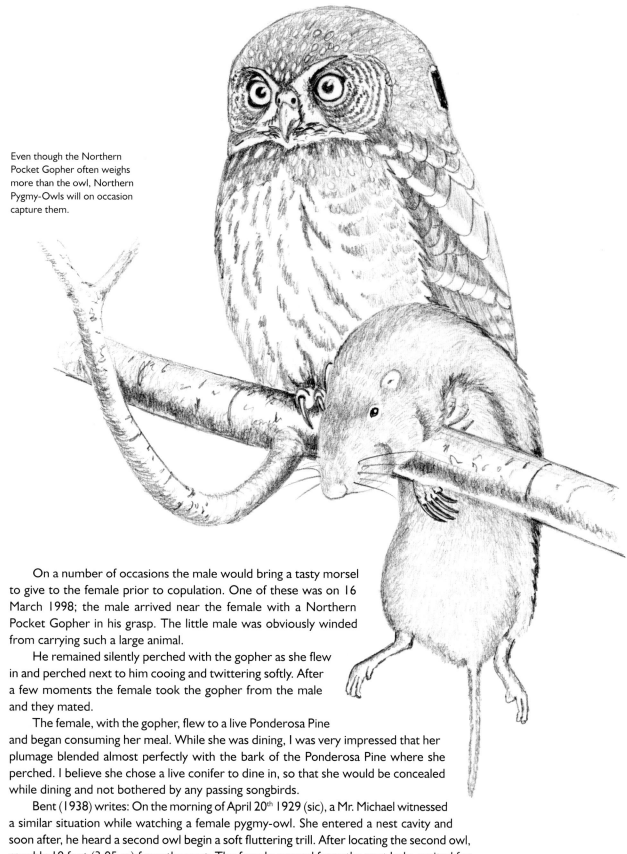

Even though the Northern Pocket Gopher often weighs more than the owl, Northern Pygmy-Owls will on occasion capture them.

On a number of occasions the male would bring a tasty morsel to give to the female prior to copulation. One of these was on 16 March 1998; the male arrived near the female with a Northern Pocket Gopher in his grasp. The little male was obviously winded from carrying such a large animal.

He remained silently perched with the gopher as she flew in and perched next to him cooing and twittering softly. After a few moments the female took the gopher from the male and they mated.

The female, with the gopher, flew to a live Ponderosa Pine and began consuming her meal. While she was dining, I was very impressed that her plumage blended almost perfectly with the bark of the Ponderosa Pine where she perched. I believe she chose a live conifer to dine in, so that she would be concealed while dining and not bothered by any passing songbirds.

Bent (1938) writes: On the morning of April 20th 1929 (sic), a Mr. Michael witnessed a similar situation while watching a female pygmy-owl. She entered a nest cavity and soon after, he heard a second owl begin a soft fluttering trill. After locating the second owl, roughly 10 feet (3.05 m) from the nest. The female peered from the nest hole, waited for a few moments before exiting the nest. She flew to the calling bird, then after a few moments the birds accomplished the supreme embrace.

Female Accepts the Male's Nest
for the Second Year

It took until the second nesting season (1999) for me to witness first hand that at times the male solicits the female while perching at the edge of the nest cavity. After hearing the tooting of the male, I moved toward his vocalizing and found him calling from the same cavity the pair nested in the previous year.

He was inside the prospective nest entrance, calling as the female watched from a few feet away. After a few minutes he exited briefly, calling with a rapid to-to-to-to-to-to-to, and the female then entered what became her home for the second straight year.

By calling and leaving food inside the nest, the male is attempting to entice his mate to accept the prospective site. Then, as she comes close to this site, he exits, calling rapidly. If she accepts the nest, his territorial tooting will diminish from that point on.

After the female has accepted the nest, she can be found within view of it for up to a week before incubation begins. Once incubation has begun, the male is seldom seen in close proximity to the nest, except to bring food to the female.

However, there were two instances in 2005 where I witnessed two different males entering two different nests after the female had begun incubating. With the first nest, the male had come near the nest with a vole and as the female exited to retrieve the animal from him, he quickly flew to the nest. He entered it just for a moment, then exited. I presume he was looking to see if there were eggs and possibly how many there were. This same action occurred with another pair in a different area that same year.

The female will remain in view of the nest until she begins incubation; presumably she incubates exclusively. However, on 19 May 2004, I was monitoring the nest and saw a very interesting thing occur. I heard an owl calling from several yards west of the nest. As I approached that owl, I noticed it had a small bird in its grasp. At the time, I surmised the bird to be either a MacGillivray's Warbler or a Common Yellowthroat. But to be honest, I could not get a good enough look to make a positive identification.

The owl with the songbird in its grasp tooted a few times, then after a short bit, a second owl came from the nest cavity and grasped the songbird from the first. Both birds were perched side-by-side about 50 feet (15.24 m) from the ground on a Ponderosa Pine limb.

Then, the owl without the songbird flew directly into the nest and remained as I watched the other owl feed on the songbird. Now, either the female came from the nest and reentered without eating, which seems highly unlikely, or possibly, both male and female Northern Pygmy-Owls incubate the eggs, but that has not been documented. Or even more bizarrely … could both birds have been female owls? The reason I suggest the latter is because that year the nest failed and until that point I hadn't had a Northern Pygmy-Owl nest fail. If both birds were females, the eggs would not have been fertilized and therefore would not have hatched.

Hunting From the Nest

From some time before egg-laying, through the first week or so after the young hatch, the male does the majority, and possibly all of the hunting for himself and his family. Holt and Norton (1986) monitored a nest where the female began hunting nine days after the young hatched. I've found a similar scenario in Colorado where the female remained inside the nest for the first 10 days or so after the eggs hatched. She spent most of the day peering out of the cavity, apparently looking for the male to bring food.

Interestingly enough, if the opportunity arises, she will hunt directly from her nest. On 22 June 1998, I arrived at the nest tree and saw her peering from her front door, obviously quite interested in something in the grass. I could hear some rustling in the grass, but couldn't identify what was there, so I remained motionless while watching the owl.

She would look in the grass and then look at me, look in the grass, then look at me. After a few minutes of this she leaped out of the nest and hit the grass feet first. I heard a squeak, then she flew to a dead snag in front of her nest with the squirming vole. She bit it in the head, looked at me momentarily and flew directly into the nest. I could hear the distinct twitting of at least three owlets inside the nest as she entered.

After the young had hatched, I showed up at the nest both morning and evening to increase my odds of witnessing interesting activity. Arriving at 7:00 a.m. on 1 July, I found the female owl perched in what seemed to be her favorite tree. It was a dead spruce about 30 yards (27.30 m) south of her nest, which gave her a good view of her homestead.

In the time I watched her, she made two attempts to get breakfast. The first was unsuccessful, but the other was quite interesting and the description follows.

After making an unsuccessful strike, she flew to a dead Ponderosa Pine slightly west of the nest. Something in a live spruce just north of her home caught her attention. I could tell she was interested in something because she was bobbing her head side to side as well as up and down. Owls' eyes are fixed in their skull, which means they are unable to move them. By moving their heads, owls are able to identify their quarry by visually comparing the potential prey to what is surrounding it.

Then after a few moments, she flew to the branch of a live spruce about 15 feet (4.5 m) from the ground, walked toward the tree trunk, and grasped something (at that point I couldn't identify it). There was a short struggle then the owl flew west, landing on a dead aspen, which is when I got a good look at her victim. It was a Western Jumping Mouse. Jumping Mice can be arboreal, meaning they are capable of climbing trees to search for food. She waited a moment, seemingly trying to catch her breath, and in "one fell swoop," flew directly into the nest with the mouse, which was larger than she was.

Northern Pygmy-Owl with a Western Jumping Mouse.

Interaction Between Neighbors

Interestingly enough, from the beginning of incubation through the day the young fledge the nest, neither male nor female ever make any attempt to capture birds near their nest. There were two instances in 1999 when I observed songbirds within a few yards of the Northern Pygmy-Owl's nest, while the female was perched nearby and she did nothing to them.

The first involved a Williamson's Sapsuckers that was passing through the area and happened to land on the owl's nest tree. When the woodpecker saw the owl perched underneath it, the owl puffed up her feathers and started snapping her bill. This was enough to cause the woodpecker to fly to an adjacent spruce and scold the owl with a series of high-pitched *PLLLLLL ----PLLLLLLL----PLLLLLL* calls for a short time before the sapsucker made a hasty retreat.

On another occasion, a House Wren was foraging directly under the perched owl. The little wren spent several minutes searching the downed logs and bushes for spiders or other tasty morsels. Yet the owl remained on her perch watching the wren. At one point the wren was within a few feet of the owl, yet the wren never noticed the little predator as the owl just watched the smaller bird. After a few minutes the wren moved out of sight.

That same nesting season (1999), a pair of Northern Flickers nested in a dead snag roughly 40 yards (36.40 m) from the owls. I was unaware of this until I heard a grouse-like sound, and I decided to investigate.

I walked several yards west to find a family of Dusky Grouse under a spruce tree. As the female grouse saw me, she let out an alarm call and her young instantly disappeared into the foli-age. She then tried to distract me by appearing to have a broken wing. Knowing I had upset her, I moved off in the same direction I came from and, while I moved back to the owl nest, I saw the female flicker fly to her nest with food for her nestlings.

I watched the flicker nest just long enough to get some great photos of both the male and female feeding their two nestlings before I returned to the owl nest. The woodpeckers fledged both of their nestlings and, again, there was no interaction between owl and woodpecker.

As I thought about it, I came to the conclusion that the female Northern Pygmy-Owl purposely left the songbirds near her nest alone because she didn't want the birds to announce the location of her nest. Enemies, knowing the location of a predator's nest, could cause harm to the young within.

It seemed very important to her that no birds or animals come near her nest at any time. I remember watching the female owl perching quietly in a dead spruce, preening and resting. The local Red Squirrel, often referred to as a chickaree, had climbed the spruce adjacent to her nest to gather cones.

Dusky Grouse are often seen within the Northern Pygmy-Owl's territory.

The owl came off her perch like a lightning bolt, hitting the squirrel with everything she had. The squirrel fell out of the tree, hit the ground, and ran back to its nest chattering the whole way. For the rest of that season, I never saw a single squirrel near the owls' nest again. For the record, there is at least one documented case of a Northern Pygmy-Owl killing a Red Squirrel, but this particular owl that I was watching apparently did not want to take it that far.

During the first two nesting seasons, a pair of Red Squirrels were the owl's neighbors. Northern Pygmy-Owls have been documented feeding upon these animals. © Steve Morello

A few days before the young fledged, the female Northern Pygmy-Owl made hunting a greater priority. I was fortunate to witness a successful hunting foray. After arriving at the nest and not seeing the female, I started searching the surrounding area until I found her several yards southwest of the nest on a dead spruce branch. She was perched on a snag looking into an adjacent live spruce. I could tell she was interested in something because she was bobbing her head side to side.

A few seconds later, she flew toward the spruce and just before she reached the tree, a female Broad-tailed Hummingbird flew out chirping and twittering. The owl was apparently after the brooding mother. But at the last moment the hummer flew off and the owl grabbed her nestling instead. The owl held the baby hummer in one foot and consumed it in two bites.

On another occasion, I was at the nest when the female exited her nest to retrieve a MacGillivray's Warbler from her mate. The owl grasped the warbler in one foot and began plucking it rapidly, making sure to turn it in such a way as to pluck its tail feathers. Then, during the owl's food preparation, I heard a small flock of American Crows moving over the nest tree. Instantly, the female stretched her neck upward and froze until the crows moved on.

After the larger birds moved out of sight, the owl relaxed and continued plucking the small bird, then ate about half of it. Then the owl walked along the branch to the trunk and wedged the remains within the bark and flew into its nest were it remained for some time.

The previous year on 3 June 1998, as I hiked to the owl nest, I found a Sharp-shinned Hawk feeding on a White-breasted Nuthatch. As the hawk saw me, it became quite aggressive and I presumed that I was close to its nest. But I made no attempt to locate it because that particular day it was a bit cold for June. I didn't want to keep the female off of her eggs.

I returned to that same area two days later with the specific intention of searching for the accipiter's nest. After just a few minutes, I located the basketball-sized nest about 18 feet (5.49 m) up in a dead spruce just a few feet off the trail. The nest was constructed over an existing squirrel nest. The female hawk had recently begun incubating and all I could see was her tail extending from the edge of the nest.

I calculated the distance between both hawk and owl nests to be roughly 350 yards (318.50 m) apart. Incidentally, the owl nest was roughly 750 yards (682.50 m) from an active Red-tailed Hawk nest as well. I found no interaction between any of the raptors.

After locating the Sharp-shinned Hawk nest and not wanting to disturb her, I rerouted my path to the Northern Pygmy-Owl nest. I would periodically check the hawk nest though to verify the date the first egg hatched. If the female was on the nest, I could pass by and nothing would happen. On the other hand, if either male or female hawk were nearby, they wouldn't hesitate to let me know I was not welcome. They would scream and dive at me, but as I passed they would retreat towards their nest.

The Sharpies eggs hatched on 13 July, but, unfortunately the nest failed on 15 July. On that day, I was on my way to the Northern Pygmy-Owl nest but noticed the female hawk acting very strange. She cackled and flew around, but made no attempt to come after me as I passed her nest.

On my way back from the owl nest, I stopped at the hawk nest to find no activity. Knowing this was a bit peculiar, I investigated the area, searching for the bird's plucking post, which is often a downed log or stump close to the nest, where the adults pluck the birds before bringing them to the nestlings. I quickly located the downed Lodge Pole Pine with several feathers and whitewash underneath. I identified several birds that the hawks were feeding on, including Pine Siskins, Downy Woodpeckers, House Wrens, an American Robin, Green-tailed Towhees, and at least one Dark-eyed Junco. To my surprise, I also found two dead Sharp-shinned Hawk chicks. I was quite surprised to find them with the remains of all the birds the hawk had fed upon. Unfortunately, I have no idea what caused the death of the young hawks.

Bringing Food to the Nest

Before the young Northern Pygmy-Owls fledged the nest each year, I watched the female owl several times bring animals to the nest and young. Some of these creatures included shrews, mice, gophers, voles, and chipmunks.

When a Northern Pygmy-Owl catches or attempts to capture prey, it will become quite excited and often twitch its tail from side to side in a rapid jerking motion.

I was present on several occasions when the male brought prey to the female who in turn gave the food to the young. But interestingly enough, I hadn't witnessed the male ever coming close to the nest or feeding the young directly.

Some of the prey items the male brought to his mate include Mountain Chickadee, Pine Siskin, Red-breasted Nuthatch, Townsend's Solitaire, and – by far the most impressive item – a Downy Woodpecker nestling.

The reason the woodpecker was so impressive (at least to me) was because it meant that the owl had to enter an active woodpecker nest to extract a bird. Mr. and Mrs. Michael (Bent 1938) tell of a pygmy-owl entering a Downy Woodpecker cavity and extracting the occupant, which was a nestling woodpecker. After witnessing the adult downy at the edge of its nest feeding a nestling and leaving, the Michaels watched an adult pygmy-owl entering the woodpecker's nest cavity and after a bit of a struggle, exited with the nestling woodpecker.

Whenever I saw the male Northern Pygmy-Owl bring in a nestling, I would picture him waiting in dense cover until his potential victim was alone, then sneaking in and taking the unsuspecting baby bird. I always wondered what the victims' mothers were thinking when they returned to their nests to find them empty.

While monitoring a Northern Pygmy-Owl nest in RMNP, Dr. Ronald Ryder witnessed, on two occasions, the male Northern Pygmy-Owl bring Spotted Sandpipers to his mate who in turn delivered them to their hungry nestlings.

The female with a Red-breasted Nuthatch that her mate has just given her. © Jim Osterberg

Food Transfers From Male to Female

I witnessed a food transfer near a nest site in 2005, that unraveled as follows. I arrived at the prospective nest area at 5:45 p.m. It was 65 degrees, overcast, and a slight wind from the northwest. I was in the small aspen grove where I had seen a Northern Pygmy-Owl previously.

At 6:15 I heard a pair of American Robins scolding something in a partially dead Ponderosa Pine roughly 150 yards (136.50 m) south of an aspen grove. As I approached the pine, I could see two American Robins calling vigorously on opposite sides of the tree with the little owl between them.

After a few moments the robins were joined by a pair of Yellow-rumped Warblers, one of which swooped down and nearly hit the owl just as a Steller's Jay joined the chorus. At 6:18 the owl flew toward the aspen grove, landing roughly 45 feet (13.72 m) up on a branch of a dead spruce.

I could see the owl had a bird grasped in its talons. Looking through my binoculars, I could see that the bird was a Dark-eyed Junco.

The owl perched on a branch of that spruce looking around and periodically calling giving three to seven well-spaced toots.

At 6:48 the female began giving her distinctive food begging twitters. At 6:57 the female flew to her mate, landing next to him. The male grasped the songbird in his bill and passes it to his mate. They mated, after which he flew to the nest, briefly peering inside, possibly looking to see if any eggs had been laid. He quickly exited and flew out of sight. The female waited a few minutes then began to consume her dinner.

Chapter Four

Egg Laying and Incubation

The number of eggs a female bird lays has a direct correlation to the amount of food the male brings to the female during courtship. The more food delivered during this time, the more eggs that will be laid (Wallace, 1963).

In a year of abundant prey, a Short-eared Owl, for example, may lay as many as eight or even 10 eggs. Yet, if there is a year where prey is in short supply, this same bird may only lay one or two eggs. If prey is almost absent, the birds may not nest at all.

With most birds of prey, the eggs are laid one or more days apart, usually in the morning. The female starts incubation as soon as the first egg is laid (Welty, 1962). The young hatch on different days and subsequently grow at different stages. In this way, if there is ever a food shortage, the older chick often gets the majority of the food and survives to fledge, while the younger chicks starve and perish.

Conversely, Northern Pygmy-Owls may actually start incubation as the last egg is laid. This way the young all hatch on or about the same day and grow at the same rate, then subsequently fledge on or about the same day. Presumably, Northern Pygmy-Owls don't have any trouble finding enough food to feed themselves and their families.

If the nest has only two young, most often, both fledge the same day, one pretty much after the other. However, if the nest has three or more young, the nestlings often fledge a day or possibly two or more days apart.

The eggs of Northern Pygmy-Owls are a short, oval-shaped, glossy white in color and about the size of a Robin's egg, averaging (depending on subspecies) 22.6 by 23.2 millimeters in length (Bent).

The average clutch size of the Northern Pygmy-Owls, at least in Colorado, is three, but they have been known to lay as many as five and even seven eggs with incubation estimated to be 28 days (Holt and Norton 1986). In Colorado, the female Northern Pygmy-Owls seem to start egg laying from mid-April to mid-May; in California from mid-April to late June; in Arizona from mid-May to late June (Bent 1938); in Montana and Oregon, from late April to late May; in Arizona from the middle of May to the middle of June (Bent 1938).

During the nesting season, the female can be differentiated from the male because the tip of her tail often gets quite ragged during nesting because it is constantly rubbing against the inside of the nest cavity while she is incubating and brooding the young.

I started thinking about the incubation period of the Northern Pygmy-Owl (approx. 28 days) as well as the amount of time the young remain in the nest (approx. 23 days) and then thought that the amount of time seemed too long compared to the size of the owl.

So, I decided to compare two birds, that were not birds of prey, one of similar size and the other of similar weight. I could not find a bird that fit both categories that was not a bird of prey. I chose the Swainson's Thrush because it is roughly the same size as a Northern Pygmy-Owl, but, unfortunately, it only weighs about half what the owl does. The thrush weighs roughly 30 grams while the owl weighs approximately 70 grams (Terres 1982). The incubation period for the thrush is between 10 and 14 days, with the young fledging between 10 and 14 days (Terres 1982).

I then looked at a bird with similar mass (weight) and chose the American Robin. The robin's incubation period is 12 to 14 days and the young fly at 14 to 16 days (Terres 1982).

I came to the conclusion that the owls' incubation period and time in the nest are so much longer than the other species because the owlets need more time to develop within the egg and more time to mature in the nest due to their ecological needs.

Nesting and Young

After the young hatch, the female will remove the egg shell fragments from the nest and drop them a few yards from it. At hatching the young are blind, mostly unfeathered, and totally dependent on their mother for protection, food, and warmth because, for the first several days of life, the nestlings are unable to regulate their own body temperature.

In Colorado, the female spends most of these early days inside the nest, only coming out when the male brings food or when she has to take care of her personal needs. In both cases she was only out of the nest for a few minutes.

Ten days or so after the owlets (young owls) hatch, they can be heard food begging from inside the nest. This call is a faint high-pitched twittering, reminiscent of the territorial calling of a Chipping Sparrow or Dark-eyed Junco.

Young Northern Pygmy-Owls having a food begging call sounding like a songbird may be a survival tactic for them, because predators would not connect that particular sound with young, inexperienced Northern Pygmy-Owls, which would be an easy meal for any would-be predator.

Two weeks after the young hatch, the female can be found perched outside the nest but is seldom more than 50 yards (45.50 m) from it, keeping it in view at all times. She can most often be seen perched in a tree that allows her direct access to the nest when necessary.

As the young begin eating on their own, she will enter the nest with food and quickly exit, leaving the youngsters to feed themselves. At this time, ants and flies tend to congregate near the nest entrance. You can often see ants and flies moving in and out of the nest.

Apparently, the flies enter the nest to lay their eggs on the rotting meat and the ants are eating both the meat and the flies' eggs.

The Owlets See More Than Each Other

In Colorado, the young can often be seen peeking out of the nest two week after they hatch and a week or two before they fledge. Often, if there are two or more chicks in the nest, two can be seen peering from the entrance at the same time.

That first nesting season (1998), I could discern three different voices inside the nest at once. When I was able to see them, I could identify the individuals by the amount of down (underdeveloped feathers) on their heads. When the owlets looked out of the nest cavity the first few times, it was so cute to watch because they would slowly climb to the entrance and look out. As they saw me, they would quickly drop out of view for a few moments, then rise up slowly and bob their heads aggressively, trying to focus on me.

After the first two days, the owlets lost interest in me. I always wondered what the little birds were thinking as they looked out of the nest the first time. Until the owlets were able to look out of their nest, the only things they saw were each other, their mom, and the inside of their nest. Then once they are strong enough to climb to the entrance and look out, everything they see must be astonishing and unbelievable to them at first.

When Northern Pygmy-Owl
nestlings are old enough to reach
the nest entrance, they often
spend most of their day viewing
their surroundings.
© Jim Osterberg

Leaving Home

In 1998, the first owlet fledged on 12 July, and the other two the following day. I could tell the nestlings were close to leaving home because for almost two weeks the little guys spent most of their day looking out their front door. They were seemingly trying to calculate the distance to the ground or the nearest safe place to land.

As I approached the nest tree, on the afternoon of the 12th, I could still hear three owlets food begging, which meant all was well on the home front. But this time, as I was just a few yards from the tree, I heard one owlet calling from outside the nest.

Interestingly enough, when vocalizing, the babies, like their parents, are quite ventriloquial. When I was just a few yards from the tree, the calling stopped, then after a little investigation, I was able to pinpoint the bird in the tree directly across from its nest perched amongst the spruce needles.

When I spotted the owl, its tiny, not yet fully developed tufts were erected, and its left wing was pulled across its breast in an attempt to remain undetected. I backed away from the nest and waited for quite a while, but after noting no activity, I took a walk around the area to check out some of the other nesting birds nearby.

There were two Northern Flicker nests a couple hundred yards from the Northern Pygmy-Owls; one that I mentioned earlier was in a dead spruce about 35 feet up (10.67 m). It had two nestlings that were about half grown. So, I moved to the second woodpecker nest a few yards farther east but found a somewhat alarming scene. Before I was close to the nest I saw that the hole had been enlarged and something was at the entrance.

When I got to it, I saw that the entrance had been aggressively enlarged and there were some Northern Flicker feathers and claw marks from a Bobcat on the sides of the tree. The cat must have climbed the tree to pull out the nest's contents. Since then, I have occasionally come across a tree or two that a cat has climbed with similar intent.

When the little owls fledge, more often than not one ends up on the ground, remaining there anywhere from a few minutes to an hour or more, resting inconspicuously as it scans the surroundings, seemingly unsure what to do next. The female will perch within a few feet of her offspring watching over them and protecting them from potential danger.

Upon returning the next day, I found all three owlets had left home. Once out of the nest, the little ones could be found, more often than not, perched conspicuously on a limb of a dead spruce, often about six or seven feet (1.83 to 2.13 m) from the ground. Even at that age, I was extremely impressed how well camouflaged the owlets were.

Their plumage matches perfectly with the color of the dead spruce and more times than not, as I was looking for them, I would hike right past one without noticing it. If the female moves into the area with food and does not immediately locate one of her youngsters, she calls softly. The young readily respond with their twittering so she can in turn locate them. Incidentally, as the birds (both adult and young) twitter, their entire body shakes. The twittering of the young is slightly higher in pitch than the adult female.

The owlets would start food begging and often continue even as the female was feeding them. The begging may actually encourage the female to continue feeding them until they're full. I remember watching the female perched between two of her fledglings, taking turns feeding both of them a chipmunk. If I had trouble locating the owlets, I would toot like the female and the young would instantly respond, allowing me to walk right to them.

Incidentally, to strengthen the owlets' wings, the female will entice her fledglings to fly to her by holding food out of their reach, which makes them come to her to get fed.

A Northern Pygmy-Owlet on its first day out of the nest trying to hide from me.

Another fledgling Northern Pygmy-Owl on the first day out of the nest. Note the short tail.

After leaving its nest for the first time, this owlet ended up on the ground. I placed it on the branch so it would be safe from ground predators. © Jim Osterberg

Fledgling Northern Pygmy-Owl, almost one week out of the nest. Note the tail has almost doubled in length since it has fledged.

Fledglings Stay Close to Home

After the young fledged that first year (1998), their mother kept them within roughly 100 square yards (84.03 sq. m.) of the nest for a little more than three weeks, before I lost track of them. During this time, both adults hunted for the family.

As the male would return with food, he would give a three-toot call. This apparently tells the female to come to him and retrieve what he has brought for her. She would go to him, taking the item, either bill to bill, or if it was a larger item, such as a chipmunk or gopher, the transfer would be from the male's foot to the female's foot. The female would in turn take the prey to the young. I've never witnessed the male ever actually feeding the fledglings.

As the young would see the female with food, they would begin begging. She would stay a few feet from them and make them come to her to be fed.

For the first few days after fledging, the female would feed the young. One of the last times I found the fledglings was three weeks after they had fledged. One was perched in a fully leafed aspen eating a chipmunk. I believe the adult killed it and delivered it to the youngster.

As I returned to my study area each day, I occasionally relocated the fledglings by either listening for their food begging, or listening for the chattering of the Goldenmantled Ground Squirrels and chipmunks. These animals seemed to continuously give up the hiding places of the owlets by chirping incessantly every time the animals saw one of the owls. The little owls were seemingly always hungry and I easily located them each day by their twittering.

I found this owlet with a partially eaten chipmunk. At that time the owlet had been out of it's nest for three weeks.

Trapping and Banding the Fledglings

As the owlets started twittering, I would often walk within a few feet of them without their getting the slightest bit anxious. Knowing I could easily do this, I decided to create a trap, enabling me to catch and band them.

I wanted to band the young birds to hopefully gain some insight into the bird's natural history. Some information I am hoping to obtain by banding the owlets includes things such as the life span of a wild Northern Pygmy-Owl, in other words, how long do they live in the wild? Whether or not the young birds return to their nesting areas to nest as adults, and if not, when the young owls mature, where do they nest?

This contraption I created to catch the owlets consisted of an 18 foot (5.49 m) aluminum extension pole with a net on the end. I developed a net-loop using copper wire and covering it with some old mist netting. The end result looked like a hand held fishing net, but the netting is much finer. It's similar to the mesh size of a hair net.

Having used mist nets for several years, I was quite familiar with how well they cling to everything they touch. So when taking the trap into the woods, the pole was in one hand and the net was in a bag in the other. My banding equipment was being schlepped by whomever I was able to talk into assisting me.

When it came time to trap the young owls, we would get close to an owlet, making sure not to frighten it. I would then place the net on the end of the pole and extend the pole so it would reach just in front and slightly below the bird. My assistant would then walk behind the owlet, and as if I had planned it, the owlet would jump directly into the net. I would quickly twist the net slightly so the bird could not escape. After removing the owlet from the net, I would put an aluminum U.S. Fish and Wildlife Service leg band on the owlets leg, check its feather condition and weight, and then place it as close to its original perch as I could before moving off.

The banding entails using a plastic leg gauge, which measures the width of the tarsus (the straight part of a bird's foot just above the toes (Terres 1982), then deciding which size band will be the best fit for the bird. I open the appropriate band, which is most often either a size 3B or 4 (butt-end) for Northern Pygmy-Owls. The band that best fits the bird is always the one to use. When placing a band on the bird's leg, you want it to fit in such a way that it can be moved around the leg as well as up and down it. Butt-end bands are so named because both ends of the band butt together securely around the bird's leg. These bands will remain on the bird throughout its life and often times longer.

When weighing the birds in the wild, I find it best to use a dark colored (blue or brown) cloth and a spring-loaded gram scale. This scale has a clamp on one end that holds the bird while the bird is in the bag. I first weigh the empty bag and then subtract that weight from the total weight of the bag when the bird is in it. This way I only get the weight of the bird.

Birds are weighed and measured in metric measurements because most individuals are too small and light to use inches and ounces and still get an accurate weight or measurement. When measuring birds, we (banders) measure the wing from the bird's wrist to its longest primary feather. Its tail is measured from its rump to its longest tail feather.

The first owlet I captured that year weighed 63 grams, had a wing measurement of 83 mm (approx. 3 in.), and a tail length of 23 mm (approx. 1.15 in.). I placed a size four aluminum U.S. Fish and Wildlife Service band on the bird's right leg. I took a few more photos and placed the young owl back on the branch where I found it.

After the owlets are trapped, I place an aluminum U.S. Fish and Wildlife Service leg band on each owlet. Then the young birds are placed back on the branch I took them from. All three photos © Jim Osterberg

The Northern Pygmy-Owl   57

The Second Nesting Season

On 26 June 1999, the female was found perched a few feet from the nest as juncos, chickadees, and kinglets scolded her. After a moment she entered the nest as the young began to food beg from within.

After two minutes, she came to the nest entrance and peered out. As she was at her front door, a Mountain Chickadee dove at her just missing as she pulled back into the cavity. As the chickadee passed, the owl instantly exited and flew into an adjacent spruce, where she sat quietly. A few Pygmy Nuthatches and a Ruby-crowned Kinglet harassed her for almost four minutes until they lost interest and moved away. This harassment consists mostly of the smaller birds flitting around the larger predator and giving some sharp chip notes. Most often during these scolding bouts the owl shows the birds little attention.

Due to the increased number of songbirds near that same nest in 1999, the female Northern Pygmy-Owl (after the young had fledged) moved her family several hundred yards east of the nest within the first few days they fledged.

Incubation began on or about 6 May, with one owlet fledging on 4 July and the other two taking the plunge on 5 July. On the fourth, I found one owlet 45 yards (40.95 m) east of the nest. The following day, all three fledglings and both adults were in the same general vicinity.

Two days later the adult female and two young were found over 300 yards (273 m) east of the nest. On 6 July they were over 450 yards (409.50 m) east of the nest. The female was able to move her family by showing them food, then forcing them to come to her to eat, and as they ate a bit, she would move away from them. If they wanted more, they had to continue coming to her.

That year I lost track of the family after 10 days. The last time I saw the owls, their mother had moved them to an area with little bird activity.

That same year, I was told of another nest roughly 15 miles (24 km) east of Estes Park in a small campground near Drake, Colorado (elevation approx. 6600 ft (2012.20 m). A fledgling Northern Pygmy-Owl was inadvertently hit and killed by a gardening tool. The owl was resting in some tall grass that the grounds keeper was grooming. He did not see the bird until he hit it.

I was called by Rick Spowart, our local Colorado Division of Wildlife officer about the bird. But as I contacted the campsite, the bird had expired. I went to check out the area the following day to find two owlets along with an adult bird in a Cottonwood tree. The only trees in the area are cottonwoods, so I assumed that their nest was in one of them.

Catching a Fledgling Off Guard

In 2000, the owls fledged three young and, fortunately, I just happened to be at the site just after the first owlet fledged. I had been photographing the nestlings for several days as they peered from their nest prior to fledging. On 25 June, I agreed to bring a birding friend, Anita, along to see the nestlings.

We drove to the parking lot and proceeded to the nest tree. We walked up the trail a short distance to the small clump of aspens, which is where the nest tree was. Then some movement left of the nest caught my eye; as I turned, I saw an adult Northern Pygmy-Owl perched on a Ponderosa Pine branch about three feet (one meter) from the ground.

The owl (presumably the female) was watching my every move; I instantly stopped and watched her for a few moments before I realized there had to be a reason she was sitting so close without moving. What was so striking to me was that she made no attempt to either attack us or fly away. She kept watching the area around the base of a nearby Ponderosa Pine. As I looked around the base of that tree I saw one of her babies on the ground.

I told Anita to keep an eye on the little guy while I went home to get my banding gear. When I returned from the 21 mile (33.87 k) drive, the bird had moved a few feet but was still on the ground. I placed a small net over the bird, picked it up, banded it, measured it, and checked the weight and feather condition, noting that the bird was in excellent condition. I then placed the bird on a Ponderosa Pine branch about six feet (1.83 m) off the ground. I took several photos of it and went home.

Ironically, the female Northern Pygmy-Owl just watched the entire time, making no attempt to attack me or, for that matter, even fly away. I returned the following day and noticed that the other two owlets had fledged and were chasing the adult through the canopy of a small grove of Ponderosa Pine directly in front of their nest. I was truly impressed with how well the young birds could maneuver in such a short time. That was the last day I was able to locate any of those birds.

Class of 2005

In 2005, after the owlets began looking out of the nest, I spent virtually every day at the site watched the nestlings as they peered from their front door. I knew the fledging date was getting closer and closer because, as I watched them, they would lean farther and farther over the edge yet never take the plunge.

On two different occasions I thought the birds had fledged and I had missed them. One of these days was 21 June. It was a sweltering 90 degrees, with no wind and subsequently no activity in or near the nest. I heard no food begging and saw very few insects at the nest entrance. I walked the general vicinity around the nest and after finding nothing I left. Upon returning two days later I found the young still in the nest. I came to the conclusion that it was just too hot for the birds to be active.

Then on 26 June, I returned to the nest tree, this time with my banding kit in hand. At 8:04 a.m. I found no activity. I did hear at least one owlet and an adult food begging just a bit northeast of me. I moved toward the twittering and saw both fledglings perched on the opposite sides of a six-foot (1.83 m) Juniper Tree.

I moved close enough to get a few good digital photos. Then I slowly maneuvered close to the owlet perched on the left side of the bush and gradually crouched down reaching out my hand and softly grasped it, lifting it from the bush. At that same time the second owlet flew straight south a few yards and landed on a low spruce limb. I made sure there was no danger near that owlet, and then proceeded to measure, weigh, and band the first.

That first owlet weighed 68 grams, had a wing measurement of 88 mm (approx. 3.5 in.), and a tail length of 28 mm (approx. 1.25 in.). I placed a size four band on its right leg, took a few more photos, and placed the young owl back on the branch I took it from.

I then walked toward the second owlet and before I was in reach, it flew a few yards away from me and landed clumsily on the ground and instantly turned its head and looked back at me. I quickly moved to the little bird and picked it up and took it back to the original juniper at which time the first owlet had flown up slope and landed next to its mom.

That second bird weighed 70 grams (2.5 oz.) had a wing length of 90 mm (approx. 3.5 in.), and a tail length of 32 mm (approx. 1.25 in). I placed the same size band on its right leg. As I was in the process of putting it on that same juniper, I heard several American Robins flying through the woods toward us, calling vigorously.

I turned to the ruckus to find an adult Cooper's Hawk flying in and landing about 50 feet (15.24 m) from me. The hawk roused its feathers and wagged its tail then looked around at the robins that were hollering at it.

I put the little owl behind my back so that the hawk would not see it. Then I walked toward the hawk until it flew away with several robins following. I waited till I could hear the robins off in the distance. That way I knew the hawk was no longer a potential threat to the owls. I moved back to the juniper so I could place the owl on it, but by this time the young bird was so stressed that it could not even stand. I placed it on a short stump, making sure mom could see it, and then I moved off. I returned about three hours later to find both young owls in different trees with their mother nearby. Unfortunately, that was the last time I was able to locate any individuals of that family.

If there are two owlets in the nest, they often fledge early in the morning. At fledging, if the owlets don't end up on the ground, they often end up perched very close to it.

An owlet trying to appear larger than it really is.

Northern Pygmy-Owlet in the author's hand.

Describing the Fledglings

At the time of fledging (leaving the nest for the first time), usually about 23 days after hatching (Holt and Peterson, 2000), the owlets still have a few down feathers on the tops of their heads and, at least in Colorado, are a kind of mouse gray. They also have white sideburns on the lower edges of their facial disk and a few white spots, mostly on their forehead and scapular feathers. They have a few spots on their backs at this age.

An owlet caught taking a nap. Note the pale eyelids.

They don't seem to acquire a significant number of spots on their faces, heads, and backs until late summer or fall. The day the young leave the nest, their tails are just a few millimeters long. It takes almost three weeks for their tails to grow to their full length of almost three inches (7.6 cm).

It's been suggested that there are some apparent sexual differences between the males and females. Young females seem to be larger than males and may be a reddish-brown, where males may be smaller and a Prout brown (Bent 1938).

After the young have been out of the nest for a day or two, they can often be found perching conspicuously yet remaining quiet. When they feel threatened in any way, they will attempt to conceal themselves by pulling a wing over their breast, squint their eyes, erect the rictle bristles around their bill and raise their feathered horns on the upper corners of their facial disk.

Over the years, I have come across several fledgling pygmies and each time they act differently. Some remain perched as I approach within a few feet and just watch me, seemingly uninterested. Yet others try to conceal themselves and some have just flown away as I approached.

That first nesting season, I remember coming upon one of the youngsters that had closed its eyes and fallen asleep while I was watching it. The young owl would look forward and close its eyes for several minutes at a time, revealing the off white colored eyelids.

I have often wondered if birds have evolved with pale eyelids so when they are asleep, they would appear to have their eyes wide open instead. Therefore a predator might think the birds are looking at them and will not attack. If you've ever watched any of the black headed gulls i.e. Franklin's or Bonaparte's sleeping, you'd have seen these pale eyelids.

Young owls are dependent on their parents for food and protection for weeks after fledging however, according to Bent (1938). Soon after leaving the nest, the young probably are taught by their parents. They learn to catch grasshoppers and from that point on their hunting gradually extends to larger game, as the young owls get older, stronger, and more skillful.

The young may begin to show sexual maturity, at least with the European Pygmy Owls, at approximately five month of age (Bergmann and Ganso 1965, Mikkola 1983).

Unfortunately, little is known about the longevity of Northern Pygmy-Owls in the wild. According to the Bird Banding Laboratory in Laurel, Maryland, since 1914 there have been 314 Northern Pygmy-Owls banded, with only one encountered. That one encounter is not enough information to get a longevity record for the species.

Incidentally, according to the Bird Banding Laboratory, a band encounter is anytime someone can read the band number, versus a recovery, which is described as actually handling the banded bird after it had been previously banded. However, in captivity, they can live more than ten years (McKeever per. comm.).

Chapter Five

Hunting and Food Habits

The business end of Northern Pygmy-Owls consists of comparatively large and powerful feet with very long, razor sharp talons (claws) which combine to make this little owl a lethal predator.

Their disproportionately large feet enable both Northern and Ferruginous Pygmy-Owls to prey on birds and animals as large and even larger and heavier than themselves.

Northern Pygmy-Owls have been known to prey on birds as large as Steller's Jays and California Quail, and animals as large as Northern Pocket Gophers and Red Squirrels. However, they most often prey on small mammals such as mice, voles, and shrews and birds such as finches, nuthatches, warblers, and sparrows. Both Northern and Ferruginous Pygmy-Owls have been known to feed on reptiles, lizards, and insects also.

Like many owls, Northern Pygmy-Owls are generalist type hunters taking whatever seems to be most readily available at the time. Some research has suggested that the Pygmy Nuthatch may be a primary food item for these little owls (Ely and Pierce 1992). However this doesn't appear to be the case. The Pygmy Nuthatch does appear in the diet of the owls, but not in any notable significance.

The primary hunting technique implemented by Northern Pygmy-Owls seems to be a wait and pounce-type method. They will often perch for several minutes (sometimes over a half hour) in one spot surveying the area before either attempting a kill, or moving on to continue searching for a potential meal.

While exploring my study area in early March 1999, I came upon an adult Northern Pygmy-Owl perched conspicuously on an aspen branch about 25 feet (6.58 m) from the ground surveying the ochre colored meadow that was spotted with small patches of snow. The bird allowed me to walk within a few feet of it without showing any interest in me.

It was very interested in something on the ground just below its perch. The owl would look at the ground for a few moments, then look at me, watch the ground, looked at me, and so on. That went on for almost three minutes.

I made no attempt to conceal myself, but instead I stood quietly against a Ponderosa Pine tree observing the owl's activity. After what seemed like several minutes, the owl hit the ground, feet first. I heard a soft squeak and then the owl bit its victim in the back of the head and flew to a nearby aspen with what I identified as a Meadow Vole in its talons.

The owl looked at me for a moment then flew into the woods and started calling. I supposed he was calling for his mate to come and get the vole.

When attempting to capture prey, Northern Pygmy-Owls (in my opinion) have a better success ratio than most raptors because they seem to wait forever before making an attempt at a kill. This way they increase their capture ratio, and reduce any potential energy loss due to exertion without energy gain.

While hunting mammalian prey in the open, Northern Pygmy-Owls blatantly perch atop a tree or shrub, overlooking a meadow or small opening within the woods. They are often found perched sometimes as high as 60 feet (18.29 m) or more while surveying their surroundings. This type of hunting is most often witnessed at dawn or dusk, after most songbirds have moved to a safe place to rest for the evening, which insures the owl won't be harassed by songbirds while foraging.

The following is an example of this: while at the Northern Pygmy-Owl nest in 2005, the female was perched on an exposed spruce limb 35 feet (10.67 m) from the ground, obviously quite interested in something on the ground several yards from her.

She dropped like a rock toward the ground then made a quick turn and rapidly flew paralleling the ground flying about eight inches (20.30 cm) off the ground till she grabbed a vole with her talons. She then flew to a limb of a spruce landing with a squirming mammal in her talons.

It was really quite remarkable to witness this hunting technique. I was quite impressed how agile the little bird was and how it flew so rapidly just a few inches off the ground, presumably so the vole wouldn't see it coming. If the prey sees the predator attacking, it will take evasive maneuvers so as not to be captured.

When hunting birds, Northern Pygmy-Owls often hedge hop, which is a method of chasing songbirds from bush to bush in an attempt to catch one. This technique can be quite effective in early summer when there are a large number of inexperienced fledglings around.

Young birds are easy targets because soon after fledging, they don't know when or where to hide when danger approaches. Hedge hopping is a hunting technique often practiced by accipiters as well as pygmy-owls.

A few hunting techniques of the Ferruginous Pygmy-Owl were documented in and around the Tucson Basin in 1996 by Richardson et al. They documented that they observed two hunting and prey pursuit strategies.

One seemed to be a kind of simple drop or pounce onto an unsuspecting animal. These attacks most often occurred from between 3 and 10 feet (.9 to 2-3 m) above ground.

The other technique was a rapid flight pursuit of roughly between 30 and 100 feet (9 to 30 m) at which time the owl would grasp its victim from its perch, which was more often than not, on the outer edge of the branches. The owls would attack at the same level as their prey or slightly below. If the first attempt failed, the chase would be terminated.

Songbirds Become Bold Around the Owl

Songbirds can become quite bold when the Northern Pygmy-Owl has prey in its talons. My guess is that songbirds know that an owl, or any raptor, with prey in its grasp, is not going to release its prey just to chase a bird.

Birds seldom harass a Northern Pygmy-Owl for more than a few minutes once the owl has prey in its talons. They seem to realize that the owl isn't much of a threat and more often than not, the birds fly way and let the owl feed. Or the owl flies to a place where the songbirds aren't going to follow, allowing it to feed in peace.

I've seen, on several occasions, this same harassment with other raptors. One afternoon, I heard several crows calling very angrily in my backyard. As I looked out my sliding glass door, I saw a Red-tailed Hawk grasping a lifeless American Crow by the neck. The hawk was perched in a large Ponderosa Pine with what I estimated as about 35 crows vocalizing in protest for several minutes before the hawk flew off with crows in tow.

Another was a Cooper's Hawk carrying a chipmunk back to his nest. The hawk was mobbed by Common Grackles as it flew through the woods back to its mate.

I've also witnessed Red-winged Blackbirds harassing a Northern Goshawk as it flew over the woods with a ground squirrel in its feet .

Filling the Pantry with Persuasion

Like other owls, Northern Pygmy-Owls will cache, or store excess food. I've witnessed this on a few occasions. The first time was during the latter part of the nesting season in 1998. The male delivered an adult Mountain Chickadee to the female. She took the songbird in her bill and immediately flew to a live spruce a few yards from her nest and wedged it between two branches.

Conversely, if the female believes her mate is not bringing enough food to the family, she'll persuade him by physically letting him know that her family would like more to eat.

It was the second nesting season when I watched this. Both adult birds were perched in different trees near the nest, apparently resting. Then the female flew directly at her mate and knocked the smaller male

Voles are a common prey item of Northern Pygmy-Owls.

Northern Pygmy-Owls are capable of capturing birds as large as Steller's Jays. This was witnessed near a bird feeder in Northern Colorado.

off his perch, after which he flew off. About 15 minutes later, he returned with a vole, which he quickly passed to her.

According to Earhart and Johnson (1970), female Northern Pygmy-Owls, on an annual basis, prey significantly more on mammals and insects than males, yet males subsequently feed more on birds than females.

Some estimates suggest that during the breeding season, the Northern Pygmy-Owls' diet may include up to 30% insects (Bull et al. 1987 and Cous 1874). Owls collected at Fort Whipple, Arizona, had fragments of grasshoppers and beetles in their stomachs, some scarcely altered by digestion. The two birds were collected before noon (Cous 1874), which shows the owls' preference for hunting during the day.

One of the more interesting items on the Northern Pygmy-Owl's menu is the California Quail documented by Thomas Balgooyen in 1968.

On 27 July, he noticed a dog playfully chasing a Northern Pygmy-Owl, which had a young California Quail by the neck. He picked up the quail while the owl was still attached. He proceeded to take both owl and quail to his house, which was 45 minutes away. At no time did the owl release its grip on the quail.

He was able to take the quail from the Northern Pygmy-Owl, examine it, weigh it (it weighed 119.0 grams), then weigh the owl (which weighed 52.0 grams), finding the quail weighed 2.3 times what the owl weighed. Both quail and owl were taken outside. After the quail was placed on the ground, the owl instantly pounced on it.

The similar Ferruginous Pygmy-Owl found in southern Arizona and south Texas, has been documented preying on birds as large as young chickens, American Robins (Bent 1938, Johnsgard 1988), Mourning Doves, Northern Mockingbirds, and Eastern Meadowlarks, and mammals including Hispid Cotton Rats (Proudfoot and Beasom 1997), Texas Kangaroo Rats, and Northern Pygmy Mice, not to mention reptiles such as Texas Horned Lizards, Six Lined Racerunners, etc.

Like the Northern Pygmy-Owl, the Ferruginous Pygmy-Owl is considered a generalist predator as well. Both species seem to feed on whatever appears to be most prevalent at that time.

Trapping and Banding the Adult Owls

During my first year of research, I made a conscious decision not to trap and band the adult birds until after the young had hatched. Also, I didn't want to trap the adult birds at the nest itself because I didn't want them to have any negative feelings about entering or exiting their nest.

I decided to combine both the bal-cha-tri and a mist net. A bal-cha-tri, or B-C, is a domed wire cage made of hardware cloth. The top of the trap is covered with nylon nooses made from fishing line and a live mouse is placed inside (the mouse never gets hurt). A mist net is a nylon mesh net, similar to a hair net. The net stretches 20 or 30 feet (6.10 m or 9.15 m) and has five pockets that stretch the length of the net. As a bird flies into the net it falls into one of the pockets and is captured.

I placed the mist net within a few feet of the nest tree hoping to catch the owls. The net was set up in a "V" shape leaving one side open and the B-C, complete with a live mouse, was placed on the ground near the center of the "V". When the owl would fly to the B-C, it would either hit the net and fall into one of the pockets or fly in the open side of the net and as it swerved away, it would hit the inside of the net and be captured in a pocket as well. The third option would be for the owl to dance on top the B-C and get caught by the leg.

With this method, I'd have several chances to catch an owl. So far this trapping method hasn't failed for me.

I capture lots of songbirds each year using mist nets, but in the area I trap, there are lots of chipmunks and ground squirrels. So I'm used to making sure the net is at least a foot off the ground so the animals can

Adult owl in hand. Note the comparatively large yellow feet of the adult bird.
© Susan Rashid

run under it without getting tangled in it. Having mist netted birds since 1994; I was quite used to keeping the net above ground. What I learned very rapidly was that as the Northern Pygmy-Owl comes to attack a ground animal, it parallels the ground and flies just a few inches above it. The first time I set the mist net, I left a foot high gap between the bottom of the net and the ground.

When the owl came in for the attack, it flew right under the net taking a swipe at the mouse and landed in a tree near the trap and watched for several minutes. I quickly lowered the net so it touched the ground. A few minutes later, the owl returned and was caught in the net. After retrieving the owl from the net, it was banded, measured, weighed, and released.

Northern Pygmy-Owls' Pellets

As you study birds of prey, you learn very quickly, that you aren't going to see every kill they make, and/or everything they feed on. But hawks, falcons, and owls and a few other species such as shrikes, kingfishers, crows, etc. regurgitate the indigestible portions of their food, including bones, scales, feathers, and fur into compact masses called pellets.

Hawk, eagle, and falcon pellets have fewer bones than owl pellets, because those species digest the majority of the bones of their prey and owls do not.

Pellets are usually regurgitated a few hours after the bird has eaten and sometimes they are regurgitated prior to consuming a new meal. When dissecting pellets it is possible to identify the items that the birds have fed upon.

However, like so many other things about Northern Pygmy-Owls, identifying prey through pellet analysis can, at times, be a bit troublesome. Northern Pygmy-Owls often take very small bites of their prey, discarding most, if not all, of its bones. Therefore, identifying what pygmy-owls feed on using pellet dissection isn't as easy as it is with larger owls. The larger owls often swallow prey like mice and other small items whole and later regurgitate a pellet that contains most if not all of the bones of their victims intact.

Some Northern Pygmy-Owl pellets were collected by Dr. Ronald Ryder, Dept. of Biology at Colorado State University in Fort Collins, Colorado, and sent to Professor David Armstrong at the University of Colorado in Boulder, Colorado, for dissection and analysis.

After the pellets were examined, Professor Armstrong sent a letter to Dr. Ryder and an excerpt from the letter is as follows: "(29, May 1986). Each pellet was dissected separately and is enclosed in a separate vial with a brief note of the obvious mammalian contents. These are a far cry from the owl pellets that I have looked at before (from barn, long-eared, and screech owls). These pygmy pellets are tiny and the owls do a real number on their food remains."

"I was surprised to find no skulls at all, only mandibles and optic capsules. There were some shrew mandibles (probably *Sorex montanus*, based on ecological considerations, but determinable only by genus), a mandible of (*Microbus*, probably *Microbus longicaudus*, with the above caveat), and a mandible of (*Peromyscus*), and each sample of teeth was associated with a wad of hair mostly one small mammal per pellet."

On 12 February 1999, I took a pygmy-owl pellet to the same Dr. Armstrong for dissection and only one small rib bone was found, but the species it came from was not identified.

To compound this food identification problem even further, in Estes Park, Colorado, we have a very large population of elk. Their droppings look very much like Northern Pygmy-Owl pellets. Just finding a pellet or two can be a bit difficult, but not impossible.

Elk droppings are circular while Northern Pygmy-Owl pellets are primarily oval and can be as large as 35 x 15 mm but average about 20 x 9 mm, which is much larger than elk droppings.

I remember watching an adult pygmy perching in her favorite tree and coughing up a pellet, which dropped directly below her. When she flew off I walked to the tree and searched underneath it for several minutes, looking over many elk duds (with a stick of course), but I was unable to find the pellet because they look so much like elk droppings.

The knife is pointing at a Northern Pygmy-Owl pellet.

Chapter Six

Northern Pygmy-Owls in Winter

Most Northern Pygmy-Owl sightings seem to occur between late fall and early spring, when the young of the year are out exploring their surroundings and adults are wandering, sometimes far from their territories.

During the winter, Northern Pygmy-Owls have been documented on the eastern plains of Colorado, several miles from their normal montane habitat. In the winter of 1999, birders noted Northern Pygmy-Owls in several locations throughout the foothills and eastern plains of Colorado. The lowest record that I'm aware of was in Fort Collins at an elevation of roughly 4800 feet (1456 meters).

Throughout the winter of 2000 several Northern Pygmy-Owls were documented in and around Estes Park, Colorado, at about 7200 feet (2184 meters).

That same year, I happened upon two different Northern Pygmy-Owls. The first was in my yard (which will be described in the next section). The other was found along the east side the local 18-hole golf course. The bird was seen along a small meadow between the course and the road.

That area has a small creek flowing through it, as well as several small shrubs interspersed, making it a decent spot for the owl to search for voles and mice.

On my way home from work, I routinely scanned the area around the golf course because it has been a good spot for wintering Northern Shrikes, American Kestrels, and other birds of prey that occasionally perch on the shrubs and power lines in that area.

On one October afternoon, while passing the area, I noticed, to my surprise, a Northern Pygmy-Owl perched on one of the many leafless shrubs. Throughout that fall and winter, several birders saw the owl almost daily, primarily in the early morning and late afternoon.

Over the last several days in December, I noticed the owl on several occasions perching on the power line just above the sidewalk. While I was out birding one afternoon, I decided to swing by the golf course to see if the owl was perched close enough to the road to allow me the chance to catch it.

As luck would have it, the owl was perched on the power line. I raced the three-and-a-half miles (5.81 km) to my home and got my B-C trap loaded with a small mouse and quickly returned to the area where the owl was.

As I came around the corner, I could see the bird still perched on the power line. I stopped my truck a few yards from the bird and slowly walked right underneath it. I placed the trap in the straw-colored grass and quickly ran back to my truck. Before I got into it, I looked back at the power line, but the owl was gone. Thinking the bird had just flown off, I slowly moved toward the trap until I noticed the owl thrashing on top of the trap.

I ran to the trap and quickly placed my hands around the little owl to secure it. I held the little bird in one hand and untangled his feet with the other. I walked back to my truck, with the bird in one hand and the trap in the other.

I banded the bird, took the regular measurements, checked its condition (fat as always), and released it. The bird flew quite rapidly across the road into the woods, at which time I lost contact with it. The following morning I didn't see the bird, yet that afternoon it was perched on one of the many bushes several yards from the road. It was still seen on and off for several weeks until courtship began, at which time the owl was no longer seen in the area.

Northern Pygmy-Owls Search for Easy Meals

Throughout most winters, and especially on overcast days, some Northern Pygmy-Owls seem to favor surveying bird feeders for potential meals. When hunting in this manner, they will remain inconspicuous while perching in a nook within a conifer or dense shrub waiting for an unsuspecting songbird, at which time the little dynamo springs into action to make the kill.

Other times, the owls will be perched late in the day, often on an exposed limb in someone's yard, searching for a vole or mouse that may be gathering seeds from under a bird feeder. There are several accounts of Northern Pygmy-Owls making kills near someone's bird feeder within view of the homeowner.

A few eyewitness accounts of winter Northern Pygmy-Owls with prey are as follows: Jack Melton of Estes Park, Colorado, told me of a Northern Pygmy-Owl he watched catch a Deer Mouse outside his house. Jack and his wife were doing dishes and noticed an owl perched on an aspen branch just outside his kitchen window.

After watching the owl for several minutes, they saw it fly down underneath Jack's bird feeder briefly, then come up with a mouse. It flew to an aspen and proceeded to consume its meal while Jack watched. During this particular hunting foray there were three-and-a-half inches of snow on the ground and it was still snowing. It was about eight o'clock in the morning.

Steven Bouricius of Peaceful Valley, Colorado, writes (*C.F.O. Journal* 1987), "At 3:40 p.m. I noticed an adult Harris's Sparrow feeding under a fir tree and as swiftly and silently as a snowball hitting a drift, the sparrow was taken down by a ball of gray feathers. The kill, made by a Northern Pygmy-Owl, was instantaneous. For a long moment there was silence. Then gradually, the scattered gathering of feeding birds assembled, surrounding the owl and its kill in a cacophony of scolding jeers, screaming about the impropriety of the intrusion. He goes on to say that the small groups of songbirds started to scold the owl and tried to drive it away, then after a few moments the owl flew to a spruce tree with its meal, where he could see that the owl was clearly smaller than its prey."

Laura MacAlister Brown from Nederland, Colorado, writes, "On Nov. 26, 2000, about 10:00 a.m., I heard a bird hit my window. When I opened the door to check it out, a Northern Pygmy-Owl flew up from the ground below the window, releasing a female Pine Grosbeak it had captured. A short while later, I saw the owl again and watched as it captured a rosy finch on the ground below the feeder. It flew up into a nearby tree and ate the finch. The owl had red phase plumage."

David Waltman, 6,000 feet (1820 meters), half way between Boulder and Lyons, Colorado, writes, "On 11 October 2001, I heard a Northern Pygmy-Owl calling in my yard. I did my owl toot, and it buzzed me a couple of times, one time within about three feet of my face. Then, it snatched a Lesser Goldfinch from my birdbath, and perched in one of my Ponderosa Pines to munch on it."

Ellen Holly Klaver from Boulder, Colorado, writes,

"It was about 4:00 p.m. on a winter's day, it was getting dark. I was walking home, in central Boulder, when I saw a movement on the ground near a shrub. When I looked closer, I saw a Northern Pygmy-Owl and a European Starling, bigger than the owl, in a grip (hand to hand combat?) with each other, struggling, on the ground.

It was a very pitched battle, but the starling finally gave out, and the owl was exhausted. I had run home to try to get a neighbor with a camera, so I missed the last gasp of the starling, but I only missed about five minutes.

The owl then proceeded to drag the starling under a bush, but s/he could only drag it one hop at a time, and then had to rest. It was very clear that the owl was using all its energy to drag the starling, because s/he had to rest after each hop, kind of drooping with the effort. But when it was under a bush the owl began eating. I was close enough to see it but not too close to disturb it. I left when it got too dark to see, and

Northern Pygmy-Owl with a
Harris's Sparrow.

Northern Pygmy-Owl with a
Brown-capped Rosy-Finch.

when I returned the next day, close to the same time, the owl was back, eating. By the following day the starling's carcass didn't have much left."

Susan Ward of Boulder, Colorado, had one of the most unusual sightings, a turquoise colored Budgerigar. It was apparently someone's pet bird that had escaped from its captor. The "budgy" had been seen in her neighborhood on and off for about two months until one day she looked out in her backyard and saw a turquoise colored bump on a branch. After examining the bump through her binoculars she saw the bump was in fact the neighborhood Budgerigar in the talons of a Northern Pygmy-Owl!

Sergeant Smith (1874) witnessed a Northern Pygmy-Owl attack a pack rat sized rodent. The mouse was on the end of a downed pine log when the little owl suddenly dropped down on it out of a pine tree about 20 feet (6.10 m) from the ground and fastened its claws into its back. The mouse ran nearly the length of the log, which was about 25 feet (7.62 m), and by the time the mouse got to the end of the log (about two minutes), the mouse was pretty well exhausted. At that time the owl killed the mouse outright. This was by no means the first rodent of this size this bird had captured (Bent 1938).

Christmas Bird Counts

Christmas bird counts have been an anticipated winter activity for bird enthusiasts since 1900 in the East and 1960 in the West. The Christmas count was started as a way to tally bird numbers in the winter as opposed to hunting birds for the Christmas dinner table.

The information acquired from these counts is a great tool for researchers to study long term trends in winter bird populations throughout North America.

Each count consists of a circle 15 miles (24.15 km) in diameter encompassing several types of terrain and habitat, which will give the optimum number of bird species and individuals counted within the allotted 24 hour period.

The participants create groups of two or more, each with an assigned area within the circle in which they count each individual bird they see. Within each state there are several count circles and the number of participants can vary from a few to several per count. The more parties within the count circle, the bigger the section of the circle that can be covered.

Northern Pygmy-Owls, being an obligate western species, have only been part of Christmas counts since 1960. Since then, the highest number of Northern Pygmy-Owls ever counted was in California in 1980 with 111 birds counted by 29 different count circles. Over the years, California birders have counted 100 or more pygmies four times since 1960, where the rest of the western U.S. counted, at the most, 34, and British Columbia, Canada, counting 70 Northern Pygmy-Owls in the same time frame. On average, California birders see 51.4 pygmies each Christmas count, British Columbia birders see an average of 20.26 pygmies, and Colorado birders see an average of 4.3 pygmies on each count.

Having been on numerous Christmas counts throughout Colorado, I have noticed that Northern Pygmy-Owls are seen on some counts some years and not on others, so I was wondering if the birds might be cyclic. So I looked at all the western states that have had Northern Pygmy-Owls on Christmas counts, but there does not appear to be any data suggesting that these birds are cyclic in any way.

An example of a cyclic species would be the Ruffed Grouse. Ruffed Grouse, in years of high food density, will essentially increase egg production and subsequently have an increase number of young grouse. As grouse numbers increase, the birds consume so much of their food supply, especially during the winter, that the bird's food supply is subsequently decreased to a level that does not give them the ability lay as many eggs the following year. Because of this limited food supply, the adults produce fewer young. Then over the next few years, with a smaller number of grouse, their food supply rebounds and this cycle begins again. (Keith 1963).

Chapter Seven

Northern Pygmy-Owls in My Yard

During the fall and winter of 2000-2001 and again in the winter of 2006-2007, I was fortunate to have two different Northern Pygmy-Owls visit my yard virtually every morning and evening. The first owl arrived late on the afternoon of 23 October 2000. At about 3:30 p.m. my neighbor stopped over and asked if I could give him a hand moving some furniture.

We went to his house and, after a short while, I walked back to my house and just as I was almost to my front door, a small bird caught my eye in the opposite neighbor's aspen tree. Without binoculars, I saw a small gray-brown bird perched on a bare aspen branch. My first thought was "that bird looks like a Northern Pygmy-Owl." But not having seen one in my yard, I figured that it had to be something else, but what?

I ran into the house to get my binoculars and looked out the front window and, to my surprise, it was a Northern Pygmy-Owl. I was so excited that I could feel my heart beating in my throat and my hands began shaking as I ran into my garage (which is attached to my house) and picked a small mouse out of my mouse house (I keep a supply of live mice on hand because I rehabilitate birds of prey and mice are the easiest creature to keep alive in the limited space I have).

I placed a white mouse inside a B-C trap and walked within 10 feet (3.05 meters) of the owl and placed the trap on the ground, making sure the owl saw it. I quickly ran back into the house as the owl rapidly flew toward the trap and briefly hovered over it before returning to its original perch, seemingly unsure what to do.

It perched in the aspen for a few moments then reenacted the same scenario. The owl perched in the tree for roughly three or four minutes before it rapidly flew over my house out of sight. I picked up the trap and went into the house. Hoping the owl was perched in the back yard; I looked out one of the back windows of my house (a few minutes later) and saw the little guy perched on a dead snag with a Northern Pocket Gopher in its talons.

It was quite a sight seeing the little owl with a gopher that I'm sure outweighed the bird. I took my camera and walked within a few yards of the owl hoping to document its remarkable capture and took a photo.

I must have been a bit too close to the bird, because it flew off as I took the second photo. The owl had such a heavy cargo that it had a very tough time carrying the animal. It appeared to take a deep breath, before taking short undulating flights from bush to bush. The bird was perched about three feet above the ground then flew a few yards almost touching the ground before landing and taking a short rest.

It was quite obvious to me that the owl's eyes were larger than his stomach. While watching the bird fly with such a disproportionately large animal, it was obvious that my offering of a small mouse was a bit puny for the bird's large appetite.

Watching him move with that gopher, it was apparent that it wasn't the first gopher that little owl had captured. I only wish I had seen the actual kill.

I have often imagined what that struggle might have been like between predator and prey. So, later that evening I decided to find out how much the gopher weighed in relationship to the owl. According to *The Audubon Society Field Guide to North American Mammals*, Northern Pocket Gophers weigh from 78 to 130 grams and conversely the adult Northern Pygmy-Owl weighs somewhere between 54 and 100 grams.

Northern Pygmy-Owls, like Great Horned Owls and American Kestrels, have the ability to carry more than their own weight.

Knowing that birds of prey often return to an area where they have had a previously successful hunting foray, I decided to wake up before daylight the following morning, to see if the little fellow had returned.

It was still dark when my alarm went off at 6:00 a.m. Yet by 6:30 it was bright enough to see the owl perched in one of my neighbor's aspens about 50 feet (15.24 m) from my house.

I went to my garage and grasped a live mouse by the tail, walked outside within about 20 feet (6.10 m) of the owl and set the mouse on the ground. After I stepped back several yards, keeping an eye on both owl and mouse, the mouse slowly moved through the straw-colored grass unaware of the owl. The little owl saw the mouse and quickly dropped from its perch swooping in like a rocket, grabbed the moving rodent, and landed in a nearby Ponderosa Pine, all in one motion.

The owl bit the twitching mouse in the back of the head, looked back at me and flew off with its breakfast. The way in which the owl caught the mouse reminded me of the way a Bald Eagle swoops low over a lake to catch a fish from its surface.

I got home from work about 3:00 p.m. that afternoon. After pulling into my driveway, I searched the trees around my house in case the owl was already in the area.

The afternoon routine at my house is to feed the cats as soon as they see the whites of my eyes, which usually takes a couple of minutes. After that, I began looking for the owl. At about 3:30, I glanced out one of the back windows and, sure enough, the little guy had returned to the same tree he was in early that morning.

I took a mouse outside and released it as I did that morning. Just as before, the owl came down grabbed the mouse and flew out of sight. This routine went on every morning and afternoon from that first evening in October through the middle of February. As the winter progressed, we had gotten some substantial snow and cold.

One morning after I released the mouse, it quickly ran into my woodpile before the owl even left its perch. So I released a second mouse, it immediately burrowed into the snow out of sight. I was able to dig into the snow and retrieved that mouse.

I took the second mouse into the house and got a towel that I took out and placed onto the snow and then released the mouse onto the towel. My hope was that the mouse would stay on the towel long enough for the owl to capture it. That way the owl would increase its chance of catching breakfast or dinner, whatever the case would be.

I even went as far as to place a light towel under a dark colored mouse and vice versa. In doing this, I would insure the owl would see the mouse. I know the owl would have seen it either way. I just loved the fact I had an owl in my yard and wanted to make sure he would return. After feeding the owl for several weeks, he had me pretty well trained. After all, how many people feed wild owls outside their house?

An interesting thing happened early one December morning. The temperature was 14 degrees below zero as the bird arrived for breakfast. Thinking he would be more hungry than normal, I decided to place my largest mouse onto the towel.

Interestingly enough, the owl showed no interest. Then after six minutes, I picked it up and put a smaller one on the towel and again, there was no response from the owl. I brought that mouse into the house and pulled out a third mouse but this time it was one of my smallest mice. I placed that little mouse on the towel and, like a bullet, the owl came out of the tree, grabbed the mouse, and flew off with it even before I got more than a few feet from the towel. At least in that instance, the larger mouse was less interesting to the owl.

For 18 mornings and afternoons, the owl swooped in, grabbed, and carried off a mouse without missing. The 19th attempt was a near miss. The owl flew down, hit the mouse, but was unable to grab and carry it off. It appeared that the miss occurred due to the angle at which he began his strike. Even though he missed his original attack, he quickly regrouped to retrieve the mouse on the second pass.

My thought is that he may have been getting so fat by that time, that he wasn't trying as hard to snatch the mice anymore.

Periodically, the owl would perch at different heights within the trees. As I released the mouse, I noticed there was a direct correlation between the angle in which he began his attack and his capture ratio.

If the bird was perched high in the tree and the mouse was directly beneath him, he would always miss the first strike, but if the angle was shallower, he was able to grasp and fly off with the mouse all in one motion.

The second owl that wintered in my yard during the winter of 2006-2007 acted quite differently than the first bird. During January 2007, this owl was both accepting my offerings as the other owl did, but this

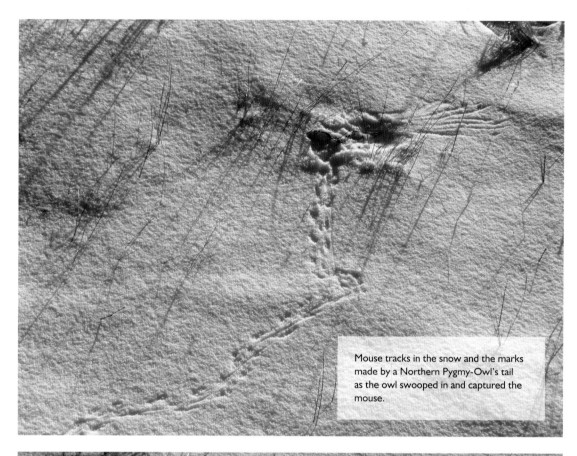

Mouse tracks in the snow and the marks made by a Northern Pygmy-Owl's tail as the owl swooped in and captured the mouse.

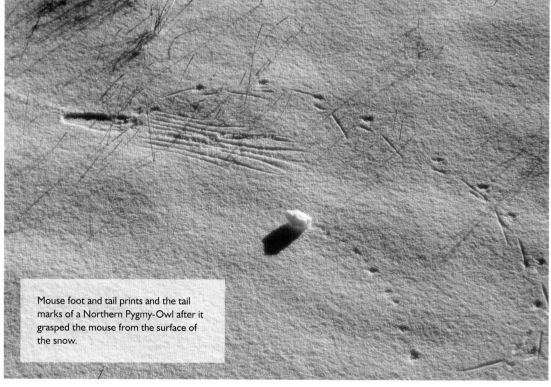

Mouse foot and tail prints and the tail marks of a Northern Pygmy-Owl after it grasped the mouse from the surface of the snow.

bird was also actively capturing birds, mice, and voles on his own. On several occasions I watched him capture House Sparrows in and around my yard. On one occasion it happened within a few feet of me.

The owl was perched in the leafless aspen in my neighbor's front yard. I had already fed him his evening meal; however, this time he had returned for seconds. As I walked out the front door, I heard some rustling in the juniper bushes in the front of our house. I kicked the bush to see what was in the bushes and a few House Sparrows flew off. Yet one was perched atop the bush, and in a split second the owl swooped in and took the sparrow from the bush and flew to the opposite neighbor's yard, landing in a Ponderosa Pine holding a lifeless sparrow by the neck. Soon after, the owl flew off out of sight.

Notifying Other Birders About the Owls

Like several birders, I belong to a birding list serve called cobirds (short for Colorado birds). It enables participants to post information on interesting bird sightings throughout the state. As the Northern Pygmy-Owl became more and more reliable, I began posting daily updates on cobirds.

It only took a few days before I started receiving calls and e-mails from interested birders wanting to see the owl. I was soon inundated with birders wanting to see this owl. I thought this was great, because I'd rather have people see this bird in my yard, under a somewhat controlled condition, than bother one during its nesting season.

Birders came from as far away as Maryland and Massachusetts to see it—well, kind of. These particular individuals were in Colorado birding already and noticed my postings on the Internet, after which they called to see if they could come and see the owl.

It got to a point that I could tell people to be at my house at between 6:45 a.m. and 7:15 a.m. and between 3:15 p.m. and 3:30 p.m. to see the bird, because he was that reliable.

Interaction Between Neighbors

During the owl's food-gathering forays, I was able to witness some first hand interactions between the owl and some of the neighborhood birds.

One of the first few days in 2000 that the owl was in my yard, it was perched in the front of my neighbor's house and a flock of House Sparrows flew from the juniper bushes in my front yard into the juniper bushes directly under the owl. The sparrows were obviously unaware of the owl.

In one motion the owl raised its little feathered horns, pulled its right wing in front of its body, erected the white eyebrows and the rictal bristles around the bill and pulled itself upright to make himself long and narrow. This was done presumably to conceal himself from the sparrows. After the sparrows were in the bush under the owl, the owl flew out of sight.

One morning after one of the Northern Pygmy-Owl's free breakfasts, he flew with his mouse to the east, but made a stop on an angled wooden post in my neighbor's backyard. At first I couldn't figure out why he stopped. So I stepped outside and noticed a small flock of American Crows flying in the direction of the owl.

In one motion the bird raised its ear tufts, pulled its wing over its prey and raised the bristles around its bill. But the most interesting thing about that incident was that the owl tilted himself to correspond with the angle of the post. He appeared to be an extension of the post so as not to be detected by the moving flock. As the corvids passed, the owl straightened up and flew off with its meal.

The Owls Become Greedy

Unlike hawks and eagles, owls and falcons store their excess food in case there's ever a food shortage. By hoarding food, they can always go back to the pantry and retrieve leftovers. This became very apparent about the middle of February 2001. The first evening I witnessed this, I was quite shocked.

As the usual time arrived, I looked out the back window just as the owl showed up in the backyard. I walked outside and released a live mouse onto the snow and he came down, grabbed the mouse, and flew off. I went into the house and, for some reason, I looked out the window and to my surprise, the owl was back in the same tree in less than 60 seconds!

I walked outside just below the owl and released a second mouse. The owl flew down grabbed it, and flew off as before. I went back into the house, looked out the back window a third time, and she was back in the tree again. I released a third mouse, and then decided that would be enough for that day, because I was running a little low on mice.

By this time, I had invested quite a bit of money in this little owl, because the mice I had were unable to breed as fast as the owl was consuming them. For the next few days the bird attempted this "returning-for-seconds" every morning and afternoon, but I caught on quickly.

Sigrid Ueblacker, president of the Birds of Prey Foundation, told me of an American Kestrel that wintered outside her facility. This little falcon learned to wait each morning for her to throw mice on the roof of her facility.

Several times throughout that winter, the little falcon would return a few times in a given morning to retrieve and store several mice that were thrown to it. As if the bird had been watching the weather channel, it would grab several mice and store them. Then as the weather became harsh, the falcon wouldn't be seen again until the weather let up.

Then the falcon would only show up for one mouse each morning, until courtship began that spring, at which time it was no longer seen.

While speaking with Kay Mc-Keever, founder of the Owl Foundation in Ontario, Canada, she informed me that in the winter her captive owls will incubate frozen mice until they are pliable enough to consume.

I know that when the owl in my yard would capture a few mice before darkness, he must have been caching (storing) them, and because of the temperature outside I'm sure the mice froze before he could consume them. Therefore he would have had to thaw his food before consuming it.

These images were created from what occurred in my backyard one morning after the owl had captured a mouse and a small flock of American Crows flew over.

Chapter Eight

Rehabilitation of
Northern Pygmy-Owls

When I began rehabilitating birds in 1994, I knew very little about what rehabilitation entailed. A great deal of my knowledge of rehabilitation was acquired at the Birds of Prey Foundation in Broomfield, Colorado. It's a wonderful facility that takes care of injured hawks, eagles, owls, and falcons. Over the years, I've become good friends with the founder and president, Sigrid Ueblacker, who has shared some of her experiences about injured Northern Pygmy-Owls with me.

One injured owl was brought to the facility unable to fly. After a short examination, it was determined that the owl had a broken humerus. The humerus is the bone that connects the elbow to the shoulder. Interestingly enough, with good food and relaxation, the bone healed in only four days, which seems quite rapid when you consider that a larger bird such as a Great Horned Owl may take two weeks or more to heal from the same injury.

One reason for the Northern Pygmy-Owls' fast recovery may be due to the smaller bird, having a much more rapid metabolism than a larger bird. The smaller bird's faster metabolism aids in its rapid recovery.

Early one spring, another Northern Pygmy-Owl was brought to the facility after it had crashed into a window with such force that it had broken off a portion of its bill.

A bird's bill is actually in two parts. The upper section is called the maxima and the lower piece (the jaw) is called the mandible. For several weeks, the bird remained in a pet carrier and, due to its injury, was fed cut up mice because it was unable to tear its food.

After a few weeks the owl was able to tear its food and feed itself, so the bird was placed in a larger flight enclosure so that it could exercise. It was checked three months later, but its bill hadn't healed well enough for the owl to be released. So, the decision was made to keep it through the winter and see what the bill looked like the following spring.

To everyone's surprise the bill grew back almost perfectly and the bird was released that spring. A bird's bill is composed of a material similar to human fingernails. We all know that our fingernails regrow after we cut them; however, neither Sigrid nor I was sure that the bill would grow back in such a way that the bird could be released. It was a first for both of us.

Another Northern Pygmy-Owl that comes to mind was one that had a small hole in its petagium. (A bird's petagium is the skin that stretches from its shoulder to its wrist.) After a few days the hole was slightly larger, and then a few days later a scab began forming. Interestingly enough, throughout the following few weeks the injury was completely healed. Birds, if given the chance, can be quite resilient and often recover from some amazing injuries.

Soon after I began researching Northern Pygmy-Owls, I realized they're relatively common throughout the area where I live. Knowing this, I wondered, at least for the first several years, why in the world I hadn't received any injured pygmies.

That mystery was solved in 2000, when at 8:00 a.m. on 30 October my pager went off and after calling the number on the pager a woman's voice explained that she found a little owl on the floor of her garage. She told me that she had walked within a few feet of it but the bird made no attempt to get away. I asked her to find a box and place a towel on the bottom of it. Then place the bird in the box and I would have my wife come and get the bird from her.

I couldn't get away from work at that moment so I called my wife, Susan, and asked if she could get the bird and bring it to me so I could identify it and check it for any injuries. About 30 minutes later Susan arrived at my work with a small box and said it was a Northern Pygmy-Owl.

I knew that Susan has spent enough time around owls to know a Northern Pygmy-Owl when she sees one. By the time Susan had arrived, I was able to get away from work. After we got home, I brought the box in the house and placed it on the dining room table. Susan brought me a small hand towel and I opened the box to see the bright yellow eyes of a Northern Pygmy-Owl looking up at me.

I slowly reached into the box with a towel in hand and gently placed it over the bird and wrapped my hand around the owl, lifting it out to examine it. The first thing I checked was whether or not the bird was fat, which is done by feeling the bird's breast bone or keel. If the bone is sharp, meaning you can feel the bone in the middle of its chest, the bird needs lots of good fresh food so it can fatten itself. If the bird is well fed, or fat, its breast feels like that of a chicken, nice and plump.

The owl was adequately plump, so I checked for any broken bones, which is done by palpating (feeling) the bones through the feathers to feel for fractures. I was happy to find nothing broken. I surmised that it was just soft tissue damage, like a strain or bruise, which most often heals perfectly and the birds can be released within a few weeks.

As I was examining the owl, Susan was setting up the pet carrier, which entails placing a layer of newspaper on the bottom of the carrier, then covering that with a dark towel and adding a shallow water dish. You want to make sure that the water dish is small enough that the bird can't drown in it. She also placed a perch (which is most often a low three-sided wooden box covered with artificial turf) and a dead mouse on top of the perch. I place a dark mouse in with any newly admitted bird because a dark mouse is more natural to them than a white one would be. Therefore, the bird is more likely to readily consume a dark mouse because wild mice are dark gray or brown. After they eat the first few dark mice in captivity, any color mouse will work.

I placed the bird in the carrier and left him alone, making sure the room was approximately 80 degrees Fahrenheit. When injured birds first come in for rehabilitation, especially during fall or winter, you need to make sure the bird is kept at a warm temperature because it reduces the odds of the bird going into shock.

To illustrate why warmth is so important, during the winter of 1996, I received a phone call from a man who had a Cooper's Hawk that had hit his sliding glass door, and was found lying on his patio. I told him to place the bird into a pet carrier making sure there was a towel on the bottom of carrier.

I then told him to place the carrier (with the bird inside) in a dark room, making sure the temperature in that room was nice and warm. I was at work and, after asking him how to get to his house, I realized that he lived just two blocks from me. I told him I would pick the bird up in a few minutes.

On my way to the man's house, I turned on the heat in my truck so the bird would remain warm until I could get it home. I arrived at his house roughly 10 minutes later to find he had placed the bird in the carrier but, unfortunately, he had placed the carrier in his garage on the cement floor. As you would imagine, the garage was quite cold. I removed the bird from his carrier and placed it into mine, then placed the carrier in my nice warm truck and drove the two blocks to my house.

I pulled into my driveway, looked into the carrier to find that the bird had expired. While the bird was cold, it was shivering and the shivering is what kept the bird alive. As I put the bird in the warm vehicle, it relaxed in such a way that it shut itself off and died. I was so angry that I wanted to call the man back and tell him how angry I was but I decided against it. I have to remember that the average person doesn't understand how important the initial care of an injured bird is.

For the rest of that day and half of the next, the Northern Pygmy-Owl, that I was discussing, hadn't eaten a thing, which is not terribly uncommon for a bird that has just been taken from the wild and placed in captivity. Occasionally this captivity can be a bit traumatic until the bird realizes it's not going to get hurt.

To entice that owl to eat, I placed two live pinkie mice in its cage. Pinkie mice are newly born mice that are too young to have any fur or even open their eyes. This type of mouse is a great thing to use when enticing birds of prey into eating because they move well enough to get the birds' attention, yet they are unable to get away from a captive bird.

The injured Northern Pygmy-Owl started eating these mice instantly and from that point on, it ate full-grown mice. One of the most important things about rehabbing birds is to make sure they have more food than they can possibly consume. All you have to do is place enough food in the cage that the bird always has leftovers. Then after a few days you can see just how much food the bird needs.

One very important part of the rehabilitation, at least for me, is to clean the bird's temporary home on a daily basis. This consists of draining and refilling the water dish, removing any leftovers that may have gone bad, then taking out the old towel and replacing it with a clean one.

To insure that the owl was not stressed, I decided to clean its cage every two days versus every day, because the little bird really didn't get the cage that dirty anyway. When cleaning its cage, I tried not to stress the owl. In other words, I preferred to clean the cage without touching the bird. The more you handle injured birds, the more stressed they become, which can prolong the healing process.

I've also noticed that as birds are in a rehabilitation situation, they learn very quickly to bite and foot anything they can get a hold of, which is most often my fingers. The term footing refers to the bird's grasping with their feet and talons. Some birds I have worked with haven't so much as scratched me, while others seem to want to cause me a great deal of harm, if they could get a hold of me. It all depends on the individual bird.

After the Northern Pygmy-Owl was in my care a few days I noticing it was getting restless. I contacted Sigrid at the Birds of Prey Foundation and asked if she had room in her facility for my little owl. She told me to bring the bird down and we would place it in a flight cage that is designed just for Northern Pygmy-Owls. It is a 10 X 12 foot (2.43m X 2.43m) wooden flight cage, three sides of which are totally enclosed and the fourth has vertical slats allowing natural light into the cage.

Inside the cage is a shallow water dish, a few nest boxes, a shelf, and several small branches that enable the injured birds to move throughout the cage. The injured birds most often begin walking and taking short flights as they strengthen their wings prior to being released. That owl made a beautiful recovery and, after several months in captivity, it was released near the protection of RMNP.

Quality of Life

One of the first things that I learned about re-habilitation is the "Quality of Life" of the individuals that are in my care. After the injured birds have gone through the needed intensive care and subsequently have spent time in large flight cages, exercising some-times for months or even years before being released, occasionally, a bird for one reason or another cannot be released. This can be caused by any number of factors. Maybe the bird doesn't see well. It may be deaf, or maybe it can't fly well enough to hunt suc-cessfully in the wild.

All rehabilitators end up with birds they can't release. It's at that point we have to decide what to do with the bird. Some of these individuals are kept and used as foster parents, raising orphaned birds. Yet others are used for educational programs. Educational programs are a great way to educate the public about birds of prey and the rehabilitation process itself.

The worst part of rehabilitation is that most of the cost of it comes out of the pocket of the person doing the rehabilitation. Unfortunately it just wouldn't be cost effective to keep each and every unrelease-able bird alive.

In injured Northern Pygmy-Owl that has nerve damage in its right wing.

Prerelease Observations

Over the years, I have released several Northern Pygmy-Owls that had been injured and subsequently successfully ameliorated. However, before the bird is freed, it needs to be evaluated to insure that it will survive in the wild. Therefore, prior to release, raptors need to be able to fly and land as well as a wild bird. They also need to hear properly, their eyesight needs to be perfect, and, for young birds, they need to know how to kill live prey and carry it off. If any one of these things is inadequate, the bird should not be released.

When Northern Pygmy-Owls are releasable, I choose to let them go at dawn or dusk. By that time of day, the majority of the songbirds either haven't awakened yet, and are still asleep, or they are settling in for the night. Either way, this allows the owl to investigate its surroundings without being harassed.

Sometimes one of these little owls comes into the rehabilitation facility in the fall but isn't quite ready for release as winter approaches. In a case like this, we will keep the individual throughout the winter and release it in the spring, a term called "over-wintering."

This is done for different reasons. For instance, the owl may not have healed properly prior to the onslaught of winter. Or the bird may be a young owl born that same year. As winter is often the hardest time for the owls to make a living, I will often keep them in captivity and release them around the beginning of March. By that time, the weather is beginning to warm up and the birds themselves are beginning courtship.

If the injured owl was an adult when it was injured, it is important to let the bird go where it was found, because an adult bird most likely has a territory and a mate that it needs to get back to. If the bird was an orphan or injured as a juvenile, it makes sense to over-winter the bird and release it the following spring in an area that is good habitat for Northern Pygmy-Owls.

Before I release a rehabilitated bird, I pull it out of its pet carrier and give it a few minutes to get orientated to its surroundings. © Susan Rashid

The Release Itself

Each rehabilitated Northern Pygmy-Owl that I have released (about 10 individuals) has begun vocalizing within a few minutes after it has flown from my hand. I assume that the bird's vocalizing is his or her version of "Can somebody please tell me where the heck I am and is it OK for me to be here?"

The day after releasing a Northern Pygmy-Owl, I return to the area just to see if I can relocate the bird and see how far it has moved. On two occasions, I was able to find the bird within a few yards of the release sight, and both times it was concealed quite well within some pine or spruce needles. It was still trying to figure out where it was and if it was all right to be there. Yet the second day after release I was routinely unable to relocate them and I always imagine that they have become successful members of the mountain community.

When I release a bird into the wild, I always think back to my first days in college. When I arrived on campus, I didn't even know where the closest tavern was, much less where my classes were, so I can empathize with those newly released birds.

Mortality

The Northern Pygmy-Owl, as with all birds of prey, is protected by state and federal law. However, that wasn't always the case and, in fact, some of these little predators were collected (shot) for private collections as well as collected in the name of research.

Over the years, Northern Pygmy-Owls have met their demise through a variety of situations. Some crash into widows primarily during winter as they hunt near bird feeders attempting to catch an unsuspecting songbird and accidentally hit a window.

The main reason birds crash into windows may be because they have no concept of windows. When they look at one, they're either seeing the reflection of what's behind them or they're seeing through the building, from one window through an opposite window, not realizing there is a building between the two windows.

Furthermore, Northern Pygmy-Owls, along with other birds of prey, occasionally get hit by cars while hunting near roadsides. At times, young owls even get attacked by larger birds, especially just after they have fledged their nest. Due to their lack of flight training, they are easy targets for avian predators.

Northern Pygmy-Owls even get attacked and killed by mobbing songbirds like jays and blackbirds. J.B. Flett wrote in 1927 that tragedy occurred in November when a Northern Pygmy-Owl alighted in a fir tree near his cottage. "I heard a battle raging outside and went out to find one owl dead and a flock of 8 to 10 Gray Jays, led by a Steller's Jay chasing another owl into the woods." (Bent 1938)

Mr. Michaels (Bent 1938) writes of a Northern Pygmy-Owl that attacked a weasel, but was apparently killed by the weasel. Mammals such as Red Squirrels and Pine Martins may be responsible for a percentage of pygmy-owl deaths, primarily when the young are in the nest (Hannah 1999).

In early February 2007, a division of wildlife officer near Littleton, Colorado, picked up a Northern Pygmy-Owl that presumably died of starvation.

While reading through the published literature, I came across a few records of other birds of prey feeding on Northern Pygmy-Owls. Mannon and Boals (1990) documented one of these little owls in a Northern Goshawk nest. Another Northern Pygmy-Owl was found consumed by a Northern Saw-whet Owl (Grove 1985). Bent wrote in 1938 about two different Spotted Owls collected in the mountains near Los Angeles that each had the remains of Northern Pygmy-Owls in their stomachs.

The similar Ferruginous Pygmy-Owl found in Arizona and Texas, like the Northern, has a few enemies as well. There are several cases of that little owl being eaten by Great Horned Owls, Harris's Hawks, and Cooper's Hawks, as well as Raccoons and Bull Snakes (Proudfoot and Johnson in press). I would think there are probably similar situations occurring with the Northern Pygmy-Owl as well.

When I release a rehabilitated Northern Pygmy-Owl, I try to release it either at dawn or dusk.

A rehabilitated Northern Pygmy-Owl a few minutes after it was released back into the wild.

Portraits of both a Northern Pygmy-Owl and a Flammulated Owl with tufts raised.

The Northern Pygmy-Owl   85

Part Two:

The Flammulated Owl

Otus flammeolus

The Flammulated Owl

Chapter One

A Flammulated Owl in Rocky Mountain National Park

Between 1994 and 2004 I had received four injured Flammulated Owls, all in the fall. Before receiving that first injured Flammulated Owl, I had just assumed that these small owls only migrated through the Estes Valley and Rocky Mountain National Park (RMNP) to nest elsewhere. Besides the injured birds that I had received, I could not find anyone who had ever seen one in the area during the spring or summer.

While reading some historical information on Flammulated Owls, I came across *Bent's Life History of North American Birds of Prey* (1938), in which it is written that one of the earliest Flammulated Owl nests documented, at least in that book, was in Estes Park, Colorado, on June 2nd 1890 (sic).

Since the creation of RMNP in 1915, no research on Flammulated Owls has been conducted within the national park, which explains why I couldn't get my hands on anything about the species from the national park.

From 1997 through 2003, I had searched areas (none too seriously, I have to admit) within the national park that I thought might be the type of habitat that Flammulated Owls would prefer. However, I was continuously unsuccessful locating any. So in 2004 I decided to find out once and for all whether or not Flammulated Owls nest in and around Rocky Mountain National Park.

At that time the biggest problem with locating Flammulated Owls, at least for me, was deciding what month to begin listening for calling males. Do the birds arrive on their territories and begin vocalizing in March, April or May? This question was answered in 2007 when sometime during the last week in April, an adult Flammulated Owl was found dead in a friend's driveway. Then, during the first week in May, I found several males calling within the national park.

According to McCallum (1994), Flammulated Owls arrive on their North America breeding grounds between April and May. However, in April the Colorado high country (above 8000 feet (2624 m) often gets a great deal of snow, so why would I look for an insectivorous owl while there is snow on the ground?

Long before I decided to search for Flammulated Owls, I knew that Common Poorwills, another insect-eating bird, can occasionally be heard calling in RMNP during April, suggesting to me that I might want to begin searching for Flammulated Owls at that time. Even though we humans think there would be no food for these insectivorous birds, they somehow find enough food to keep themselves going.

While talking with other birders about Flammulated Owls, a few told me that if I can locate Common Poorwills I would have a good chance of finding Flammulated Owls because both species are often found in the same habitat.

Earlier in the summer of 2004, I was told by some birder friends of mine that Common Poorwills were found along the Cow Creek trail on the north end of RMNP. On 26 May that year, a few of us began hiking up the trail before dark in search of Common Poorwills. As darkness fell, we found several Poorwills calling throughout the valley with one perched in front of us right on the trail itself; so I thought that area might be a good spot to begin searching for Flammulated Owls.

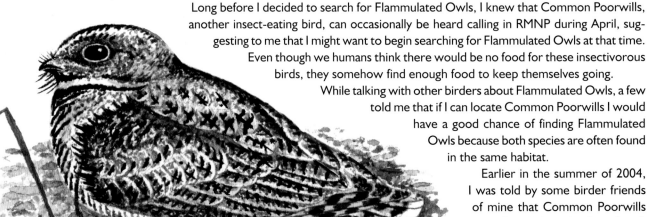

Common Poorwills are often found in the same areas as Flammulated Owls.

Seeing My First
Flammulated Owl in the Wild

Being an avid birder and part of a birding listserve called cobirds (short for Colorado Birds), I knew that Flammulated Owls regularly nest both north and south of RMNP.

Then, as luck would have it, I received an e-mail through cobirds stating that a Flammulated Owl was seen for several days in an abandoned woodpecker cavity on private property. I contacted the landowner and asked if I could come to her house with the intent to gain some insight into the habitat these owls favor.

It was about 8:00 p.m. when I arrived at Carol's house. It was cool and calm that evening. The sun had already set, yet the residual afternoon light made it quite easy to see our surroundings.

She had previously invited a few other birders who were already present by the time I had arrived. After a quick introduction, we made our way through a gate, down a short trail to the mixed tree grove at the edge of a meadow where Carol pointed out the nest tree. The vegetation on the forest floor was tall grass and shrubs interspersed between the Quaking Aspen, Ponderosa Pine, and Douglas fir trees.

I set up my tripod, spotting scope, and digital camera with the hopes of seeing and possibly photographing my first Flammulated Owl before we lost the evening light. The owls had chosen to nest in a flicker-sized cavity about 15 feet (4.57m) up in an aspen. Around the nest opening were dark warty patches, which, as I later noticed, are around most Flammulated Owl nest cavities that I've come across.

A few moments later, a Red Squirrel chattered from a tree behind the owl's nest. Instantly the female owl was perched at the nest opening looking around. She was seemingly quite concerned about the noisy mammal. The owl looked around for a few moments then backed down into its nest until about 9:00 p.m.

At that time, the male began calling a few yards from the nest and the female came to the entrance again and waited for several minutes. By then it was too dark to see clearly and we left. We all got a great look at the owl and soon after moved away and had a short discussion about how great the little owl is.

I made my way back home and while driving down the dirt road from Carol's house there was a Common Poorwill perched in the gravel alongside the road. The Nightjar was not far from the owl's nest, which reiterates what I was told about the relationship between the two species.

On the way home, I began visualizing parts of the national park that I believed would have habitat adequate to support Flammulated Owls. The first area that came to mind was the Cow Creek area that Northern Pygmy-Owls had used for nesting in 1998 and 1999 because that area has similar habitat to what I had just seen.

An Adult Flammulated Owl about to leave its nest. © Bill Schmoker

Finding a Flammulated Owl in RMNP

The following day, I decided to hike to the portion of Cow Creek that I felt might be adequate habitat for Flammulated Owls. The immediate area in question has three dominant tree types: Ponderosa Pine, Douglas fir, and Quaking Aspen, most of which have numerous cavities.

The best tree in that general vicinity for them to nest in is a dead aspen that has numerous abandoned woodpecker cavities in it. However, just a few weeks earlier, I found a Northern Flicker peering from one of the west-facing cavities that had been occupied by a pair of Northern Pygmy-Owls in 1998 and 1999. So, I expected the Northern Flicker to continue occupying that particular hole.

I tapped on the east side of the tree, hoping that a Flammulated Owl might have chosen one of the other cavities within that tree.

After tapping on the east side of the tree and seeing no activity, I moved to the west side and, to my amazement, a Flammulated Owl was looking out briefly, then quickly reentered the same nest cavity that the Northern Pygmy-Owls and Northern Flicker had previously used.

Flammulated Owl habitat.

Description of the Flammulated Owl

The Flammulated Owl is one of the smallest North America owls, measuring about seven inches (17.8 cm) from head to tail; as with most other owl species, the females are slightly larger than the males. However, there is a fair amount of overlap within the sexes, so with some pairs this size difference is not very obvious. Flammulated Owls have a vermiculated gray coloration overall, long wings, a short tail, tiny ear tufts, and dark eyes. Their dark eyes and small size make them unmistakable, considering all the other North America owls of similar size have yellow eyes and lack ear tufts.

One might mistake a Flammulated Owl for the much larger screech-owl, but the Flammulated Owl is about half the size, has much smaller ear tufts, and again has dark eyes.

Throughout their range Flammulated Owls, like Northern Pygmy-Owls and screech-owls, are found in both red and gray color phases. The red-phase birds have rufous coloration on their facial disk and scapular feathers while the gray-phase birds have a gray facial disk and a bit less rufous on the scapulars. There does not appear to be a clear-cut line that delineates where the two color phases can be found. However, the

Adult Flammulated Owl in the author's hand. Note the owl's small size and the rufous scapular feathers, which give the bird its name. © Susan Rashid

reddish birds are found in the Flammulated Owls' more southern range. These red-phased Flammulated Owls are rare within the continental United States (McCallum 1994).

The vermiculated markings of these owls helps them to remain concealed during the day when perched in dense foliage or against the trunk of a tree because the owl's feathers look remarkably like the bark of the trees they roost in.

The owl's legs are feathered to the toes, even though the toes themselves are unfeathered. Their toes are so small, in fact, that when perched on my finger, its toes only fold about halfway around it.

What's In a Name?

Originally, these birds were named the Flammulated Screech-Owl, then later the screech was dropped, which may have been due to the fact that the owl's primary call is a hoot, not a screech, although when stressed or scared these little birds will let out a soft screech-like sound.

When dissecting the birds' scientific name, *Otus flammeolus, Otus* pertains to the bird's tufts and *flammeolus*, meaning flame-colored or flaming, comes from the rufous colored scapular feathers on the birds back.

Distribution and Range

Throughout North America, the Flammulated Owl has been found from southern British Columbia south through the east slopes of the Cascades and interior areas of Washington, Oregon, northeastern California, and western Nevada. In California, the owls are found in both the Cascades and Sierra Nevada mountain ranges, as well as numerous areas throughout western Colorado, New Mexico, and Arizona, and parts of Guadalupe along with the Davis and Chisos Mountains in Texas (McCallum 1994). There are several areas throughout Mexico where these owls nest as well.

Being insectivorous forces Flammulated Owls to migrate in the fall south to Mexico to search for food, yet there are a few winter records of these owls in Arizona, Louisiana, and California (Collins, et al. 1986).

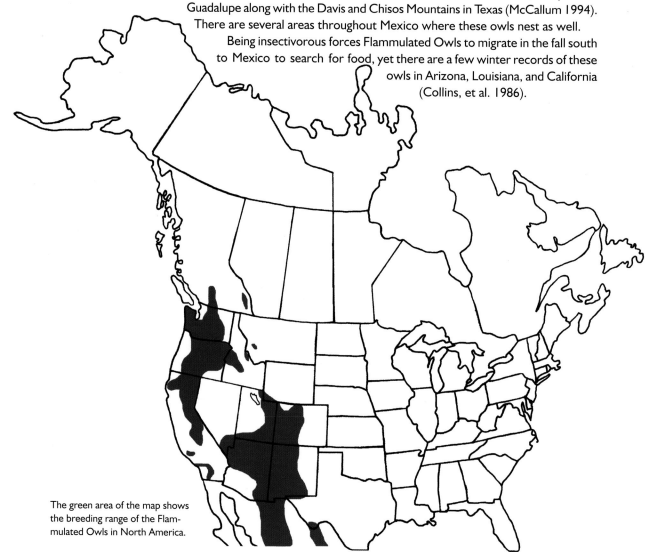

The green area of the map shows the breeding range of the Flammulated Owls in North America.

Chapter Two

My Flammulated Owl
Research Begins

In the spring of 2004, a Mountain Lion had killed an elk in RMNP, just up the trail from the nest tree that the Flammulated Owls were using. Therefore, I made sure to always bring someone with me while researching, just in case the cat showed up. I thought it might be helpful to bring someone I could outrun.

Late that first evening (12 June 2004) about 8:15 p.m., I arrived at the nest with three friends to verify the Flammulated Owl was in fact nesting in that tree. We sat within view of the nest for several minutes, yet neither saw nor heard any owls. At about 8:50 a male Flammulated Owl gave a series of soft hoots from about 40 feet (12.20 m) west of the nest tree.

I walked toward the owl, but before I reached him, he flew over my head in the direction of the nest. At the same time a second Flammulated Owl (presumably the female) exited the nest and met him. The birds were perched side by side on a spruce branch a few yards from the nest. We got a good look at both birds and verified that they were in fact Flammulated Owls in that nest. We waited a few moments hoping to see a food transfer, or something of that nature. But, the birds flew off into the darkness and, not wanting to bother them, we hiked back through the woods to our cars.

Not knowing where the birds were in their nesting progression, I contacted Brian Linkhart, a Flammulated Owl researcher, who told me that if the female is disturbed too much, specifically before the eggs have hatched, I might run the risk of her abandoning the nest. He also explained that the male will come to the nest with food every minute, or so, for the first three nights after the eggs are laid and again just after the eggs have hatched.

The male Flammulated Owl brings food with this frequency after the eggs have been laid because the female needs to replenish the calcium that was depleted from her body during the egg production process. His feeding schedule also occurs after the eggs hatch because the owlets need that kick-start of nourishment.

After seeing my first Flammulated Owl at Carol's house, I began to read a bit about the species and found that during the first few days after the eggs hatch, the female Flammulated Owl remains in the nest with the young during the day and the pair begins activity very close to 9:00 p.m. Therefore I decided that we should return to the nest the following evening and subsequent evenings about 8:30, making sure to be seated before the female exited each evening.

My hope was that if I were to sit quietly she would not find us to be a threat and would go about her normal routine.

When the Northern Pygmy-Owls used that nest, I was able to sit within a few yards of it; so I decided to try sitting in the same spot while studying the Flammulated Owls. The adult birds didn't seem to mind in the slightest.

Throughout the Flammulated Owls' nesting season, we routinely heard both Common Poorwills and Common Nighthawks calling each evening.

On two occasions one of these nightjars flew right past us while we were seated near the nest. The first bird was a Common Poorwill. It was just before 9:00 p.m. one evening as we were waiting for the evening activity to begin as the dark gray goatsucker flew between the two of us from back to front then flew out of sight.

On the second occasion, a Common Nighthawk was hawking insects above our heads and for a brief moment it flew so close to our heads that we could hear its wing beats as it passed.

As we hiked out of the owls' nesting area each evening, we were more often than not entertained by the antics of several fireflies in the tall grass of the meadow. Until that evening I was unaware that these insects were found above 6000 feet (1829 m).

The Male Provides for the Family

From the beginning of incubation, in 2004, until the owlets were better than half-grown, the male did all the hunting for himself and his family. To identify what he was bringing to the nest, I'd occasionally have to shine a light on him. Not wanting to frighten him, I placed orange cellophane over the light so when it was shined, it wouldn't be bright enough to scare him. In addition, the use of the flashlight was kept to a minimum.

In 2008, while monitoring a Flammulated Owl nest, I was able to sit close enough to the nest that I could take photos of what prey items the adults were delivering to the nestlings.

In both cases I set up my camera, complete with flash, well before dark, focusing the camera on the nest cavity and using a shutter release and flash to take the pictures. In 2004, I did not photograph the female because I wanted her to have no fear of me or the nest itself. From the time she exited her nest until after she reentered it, I took no photos. I made sure to photograph the male only, because he didn't seem to be bothered by me or the flash.

Except on evenings when it was overcast, the owls' routine began each evening about 9:00 p.m. On overcast evenings, the male would begin his evening routine closer to 8:30. He would give his distinctive three-note *hoot*, and then the female would perch at the nest entrance momentarily before exiting. She would, more often than not, spend less than 15 minutes out of the nest before reentering and remaining there for the next several hours. Keep in mind that we only stayed in the area until 11 or 12:00 p.m. each evening, so she could have exited the nest after we were already gone.

After the eggs hatched in 2004, the male would appear near the nest with food in his bill. Occasionally, we could hear the female giving a soft hoot while the male was hunting. Prior to entering the nest, the male would hoot, wait a moment, then enter. It appeared that he called to his mate so she would know he was about to come through the front door.

What I found so amazing was the frequency with which he would return with food. On 19 June, I documented the following: At 9:07 p.m. the male arrived at the nest with a small moth in his mouth. He entered the nest and remained in it for less than 10 seconds, then exited headfirst. At 9:09 the male returned with a moth, 9:10 he returns again, 9:12 same, 9:14 same, 9:17 same, 9:18 same, etc. The male would return with a moth, beetle, or other item almost every minute until we left that evening. However, according to Reynolds and Linkhart (1987) these frequent food deliveries do not continue all night. I was very impressed with how the owl would exit the nest, locate a prey item, catch it, and return to the nest, all within roughly 60 seconds. It also told me that he must have been hunting within about 100 yards (91 m) of the nest to be able to return with this frequency.

In 2008, both adult birds began feeding the owlets within a few days after the eggs' hatching. Therefore, this pair was much less vocal. The male would hoot a few times just at dark. The female would peer from the nest, then after a few minutes, she would exit and both adult birds would begin feeding the owlets in the nest. That year there was virtually no vocalizing by the adults after the female exited the nest.

Adult Flammulated Owls search for food more often in old Ponderosa Pine and Douglas fir trees than other tree types within their territories; they also frequently use these tree types for day-roosting and resting as well (Linkhart et al. 1998). In some areas of the owls' range, Idaho (Powers et al. 1996) and southern British Columbia for example, (Howie and Ritcey 1987), Flammulated Owls will nest in Douglas fir trees when no Ponderosa Pines are present (Linkhart et al. 1998).

According to Furniss and Carolin (1977), Ponderosa Pine and Douglas fir trees contain up to four times as many lepidopteran species as other western trees. Lepidopterans include moths, and butterflies.

Adult Flammulated Owl
returning to the nest with
a moth.

The same bird exiting the
nest a few seconds later.

The Owl's Neighbors

One evening in 2004, while waiting for the owls to show themselves, we heard, from several yards east of the nest, a flock of scolding American Robins flying toward us. A few moments later, a Great Horned Owl landed a few yards from the nest with several robins voicing their disapproval of the avian predator.

The robins landed all around the large owl, vocalizing loudly. A few minutes later, the owl saw us, then flew west with the robins following. During all this, the female Flammulated Owl must have exited the nest, because a short time later we could hear both male and female Flammulated Owls calling several yards west of the nest.

The next interaction occurred on 22 June: the male began calling well before dark and instantly a House Wren heard the owl and began harassing it verbally until just a little after 8:50 p.m. During the wren's vocalizing, the owl was silent. At 8:50, when the wren stopped calling, it must have moved off to roost for the evening. A few minutes later, the owl began hooting and the evening routine continued.

Another interesting interaction came about on 16 July. We were at the nest at our normal time as a Northern Pygmy-Owl began tooting just southwest of the nest. I walked to the pygmy-owl and found it perched on a dead spruce branch giving its distinctive toots.

After a few minutes it flew east until it was pretty much straight south of the Flammulated Owl nest. After about a minute, the Flammulated Owl began calling. The Northern Pygmy-Owl flew down in the direction of the calling Flammulated Owl and perched about 40 yards (12.19 m) west of the Flammulated Owl nest and continued calling. It was really interesting to hear these two species calling simultaneously with no interaction between them. A little after 9:00 p.m., the Northern Pygmy-Owl quit calling and the Flammulated Owls continued their evening ritual.

In 2008, there were very few birds in the area of the Flammulated Owl nest. Therefore, I noted no interactions between species.

Hunting and Food Habits

These owls are, for the most part, completely insectivorous, hunting primarily after dark and just prior to sunrise, with less activity during the middle of the night (Marshall 1957).

Interestingly enough, there is at least one account of a researcher finding feathers of a junco in a nest (Bull and Anderson 1978). However, it's more likely that the junco was a leftover from the previous tenant.

During nesting, I've witnessed the adults bringing moths, beetles, larvae, and/or caterpillars to their nests. The largest creature that I watched the male deliver to the nest was possibly a Columbia Silk Moth. Other researchers have documented grasshoppers, crickets, spiders, scorpions, beetles, centipedes, and millipedes.

Like the Northern Pygmy-Owl, Flammulated Owls have virtually symmetrical ear openings, suggesting they locate their prey by visual, rather than auditory, cues.

The hunting techniques of the species vary depending on where the prey is before it is captured. I've seen Flammulated Owls hawking insects from the edge of an evergreen bough, seemingly hovering at the limb edge and daintily grasping the insect with its feet before flying to a perch where it transfers the prey to its bill. I've also watched the male fly into the grass to capture prey as well.

While seated near the nest one evening, the male returned with a beetle and delivered it to his mate inside the nest. He then quickly exited, landing on a spruce bough near the nest, waiting momentarily as something above the nest caught his attention. He flew to the tree, contorting his body in such a way that, for a moment, it made me wonder if he were made of rubber. He then grasped the insect with his toes and returned to the original branch momentarily while transferring the insect to his bill before entering the nest.

The summer of 2004 was an unusually wet one; due to the rain, there were several evenings when my research was either cut short or never began. However, there were a few evenings when my assistants and I sat through a light rain just to see what the owls would do. During those rainy evenings, while the male hunted, his feathers became quite wet and as he flew, we could very clearly hear his wing beats. On a few instances, after the male exited the nest, he would fly directly over my head and his flight made a great deal of sound for an owl. However, while his feathers were dry, I couldn't hear his wings beating at all.

Flammulated Owl skull. As with Northern Pygmy-Owls, Flammulated Owls have symmetrical ear openings.

Close-up of Flammulated Owl wing edge. These owls do not have a completely silent flight

Pellets

The pellets that these owls cast are usually quite small and more round than oval as the larger owls' pellets are. Of the few pellets that I've retrieved, they averaged 13.4 x 8.6 mm and contained mostly insect remains.

Nesting

Like the Northern Pygmy-Owls, Flammulated Owls are secondary cavity nesters and seem to prefer nesting in abandoned cavities excavated by Northern Flickers. Along with flicker cavities, Flammulated Owls have been documented using natural cavities and occasionally nest boxes as well (Hasenyager et al. 1979).

Typical Flammulated Owl habitat.

Female Flammulated Owl looking out of her nest just after dark.

As for tree preference, Flammulated Owls have been found nesting in aspens and Cottonwood trees, pinion, oak, and Ponderosa Pine, (McCallum and Gehlbach 1988). In Northwest California, Marcot and Hill (1981) found the birds nesting in areas with California Black Oak and Yellow Pine. In Arizona, they are found in mixed Ponderosa Pine and oak forests (Phillips et al. 1964). In northern Colorado, the preferred habitat appears to be a mixture of Ponderosa Pine, aspen, juniper, and Douglas fir.

Of the six nests that I've come across thus far, all have been in aspens at least 8,000 feet (2,439 m) above sea level (two were at least 8,240 feet (2,512 m), with Northern Flickers believed to be the architects of each nest.

Five of these nests were in cavities 16 feet (4.88 m) or higher (one nest was over 30 feet (9.15 m) from the ground) and each cavity had black warty patches around the entrance. Having checked several hundred nest cavities over the years, and finding the majority of Flammulated Owls nesting in the aspen trees with these black patches around the entrances makes me wonder if this type of cavity might be preferred by the owls.

Maybe the owls search out these warty entrances on purpose because it aids in the concealment of the female when she's at the nest entrance. If and when she's perched at the nest entrance, specifically during the day, she often squints her eyes and raises her ear tufts and the rictal bristles around her bill in an attempt to make herself appear like a knot of the tree instead of an owl and hoping she will be overlooked by any would-be intruders.

During the summer of 2008, I located a Flammulated Owl nest in and aspen that was eight feet, seven-and-a-half inches (2.43 m. 18.41 cm) from the ground. This was the lowest Flammulated Owl nest that I had found up to that point. Furthermore, that cavity entrance had no black warty edges to it, which was also a first for me.

That cavity had an entrance size that was two and three-quarters by two and three-eighths inches (7 cm. x 6.6 cm). However, a few inches above the nest cavity was a large black patch. When the adult was perched at the nest entrance, you would notice the large patch above the owl before you would see the bird.

I made other observations during these nesting seasons that included the vegetation around the nests. This vegetation consisted of various grasses, some of which were knee high or higher, and shrubs such as Wax Currant and Common Juniper along with downed logs of various sizes.

The trees in the areas are aspen, Douglas fir, Ponderosa Pine, poplar, and juniper as well as several dead trees, most of which have several nest cavities in them. Also within these areas is an active water source. Apparently Flammulated Owls choose this mixture of trees and vegetation, which allow food, cover, and nesting sites.

Another interesting observation made during the 2004 nesting season was that most of the nest cavities that the owls chose had a branch or branches on or within a few feet of the nest itself, allowing the male to land and often call near the nest briefly before entering the cavity. It appeared important to him to land on a branch close to the nest briefly before entering. Between 95-99% of the time that I watched the male return to the nest with food, he would land on one of these branches prior to entering. It appeared that he was landing on a branch so he could judge the distance from perch to the nest before entering.

However, in 2008, when both male and female were hunting, the pair would most often either fly directly into the nest without stopping, or perch at the nest entrance momentarily before entering to feed the owlets.

When both male and female are hunting for the family, there is no need for the male to perch near the nest before entering. Therefore, he and his mate would fly directly into the nest to feed the young. The male would not need to land near the entrance to let the female know he was coming in because she was already out hunting too.

From top:
Flammulated Owl nest tree.

View from the owl's front door.

During the 2008 nesting season, both male and female fed the nestlings. This photo shows one adult at the nest and the other flying nearby waiting to enter the nest.

On the evening of 7 July, I set a trap to catch and band the adult birds at their nest. The trap was a mist net placed in a "V" with the bottom of the "V" a few feet in front of their front door. A few minutes after 9:00 p.m., the male was caught as he attempted to enter the nest. Soon after that, the female was caught, as she attempted to reenter the nest. Both birds were measured, weighed, banded, and released. The male had a wing length of 135.5 millimeters and weighed 50.2 grams while the female's wing measured 137 millimeters and she weighed 59.5 grams.

As mentioned earlier, Flammulated Owls occasionally nest in nest boxes to raise their families. So, if you put up boxes for them or for other small owls, keep in mind that you need to place straw, grass, or pine needles in the bottom of the box. This way the eggs won't roll around as the female enters and leaves the nest. The bedding will also insulate the eggs so they won't chill from below.

Adult owl in the author's hand: note the
bird's small size and the leg band.
© Steve Morello

Adult male Flammulated Owl after
being banded by the author.
© Steve Morello

Territory Size

While watching the frequency at which the owls delivered food to the nest and young, I surmised that the birds' territory must be rather small. However, after reading some telemetry research done on Flammulated Owls in Colorado, I found they can have a quite extensive territory for such a small bird. In fact the owls' territory range is from 8.5 to 24.0 hectares (21 to 59.30 acres) (Linkhart et al. 1998).

However, the owls have what is termed intensive foraging areas, or IFA, within their territories. This IFA is a kind of smaller micro-territory within the territory that the male uses to locate and capture food with more frequency than he uses the rest of his territory. Some individuals have more than one of these areas within their territories (Linkhart et al. 1998).

Also within the territory, the birds have particular trees that they often call/sing from. When vocalizing, they're often within the lower sections of the trees near the trunk. My guess is that this is done for security purposes. In other words, they call from a concealed area to protect themselves from larger nocturnal predators. They will at times vocalize while perched in the open as well (Kevin Cook per comm.)

During the day, the males most often roost in the concealment of a conifer such as a Douglas fir or spruce tree. They typically pick a roost tree that has a dense cover of overhanging horizontal limbs and/or foliage (Linkhart et al. 1998). I have found these owls roosting on several occasions and each time in either a Douglas fir or Quaking Aspen. The birds roosting within the Douglas fir trees were adults and in each case the birds were perched tightly against the trunk of the tree surrounded by branches to aid with their concealment.

The one bird I found roosting in an aspen was a fledgling that was discovered by some mobbing warblers and chickadees. That bird was not at all concealed, but after the birds harassed it for a few minutes and moved on, the owl remained resting in that tree until it moved off after dark.

Adult males appear to be much more faithful to the territory than the females. According to Reynolds and Linkhart (1986), male Flammulated Owls remained faithful to their breeding territory once it is established. However, some females will change both territories and mates from one year to the next.

As with most migratory birds, female Flammulated Owls arrive on their breeding grounds after males, and subsequently have the option of choosing a mate with the best territory. However, the females do not appear to assist the males with any territorial defense (Reynolds and Linkhart 1986).

Young, inexperienced males attempting to establish their first territory may be forced to a less adequate area and remain unmated.

Chapter Three

Vocalization

As spring arrives, the male Flammulated Owl can be heard vocalizing within his territory with the intention of attracting a female. When hooting, he often calls from a concealed perch, enabling him to advertise without running the risk of being attacked and possibly killed by a larger owl.

The advertising call of the male Flammulated Owl is a series of very low frequency hoots, described by Marshall (1939) as about B above middle C. A calling male often begins hooting softly and at a low pitch, then gradually intensifies until the tone is reached at B.

This call is a three-note *hoot* given in a sequence of *hoot,hoot........hoot ....hoot,hoot............hoot*, sometimes described as *boot,oop......poop..........boot,oop.......poop*. It can on occasion be varied to a kind of....*boot,poop;bootle,poop;bootle-oop,poop* as well (Marshall 1939).

The hoots uttered by the male are much lower in pitch than the female, whose higher pitched single *hoot* is often heard in response to his calls.

Locating a calling Flammulated Owl can be quite troublesome at times because its voice doesn't seem to carry more than a few hundred yards, which may be due in part, to the dense foliage that the birds vocalize from, or perhaps, it may be that the call itself is quite soft and low in pitch.

A second problem occasionally occurs because the owl's voice often has a ventriloquial quality to it, making the calling birds seem to be in a much different place than they actually are.

As the males arrive on their territories in the spring, they can be, and often are, quite vocal, calling for a prospective female throughout the evening. Marshall (1939) found on 20 May, an individual male calling almost continuously throughout the evening, but paused its calling every 15 to 25 hoots seemingly to listen for a response. The bird was also silent for several minutes as it moved from one perch to another. It was also silent for several minutes prior to its final calling bout before sunrise.

Occasionally, paired birds' territories will overlap slightly showing how tolerant they are. When males from adjacent territories end up within the same general area, a protest call will often be uttered, but there does not appear to be any physical confrontation.

When a local male becomes agitated due to an intruding male within his territory, the territorial birds' three-note hoots will become a bit more rapid and evenly spaced with a sequence of *hoot,hoot,hoot*. Marshall described it as a hoarse, breathy, rushing sound, similar to that of the courtship dive of a nighthawk.

Like other owls, these little guys seem to reduce the intensity of their territorial vocalizing after the eggs have been laid. Often, when the male comes to the nest with food, he perches near it and utters a soft three-note *hoot..hoot....hoot* prior to entering the nest.

I've noticed, on two different occasions, that the male will cease hooting if a larger owl such as a Great Horned Owl either begins calling (in relative close proximity), or if the Flammulated Owl is vocalizing as a larger owl moves in and is either seen or heard by the Flammulated Owl. Marshall found a similar occurrence when a Spotted Owl was calling near a Flammulated Owl. The smaller owl would freeze and remain quiet when the larger owl was heard.

The distress call of this little owl is a kind of quavering whinny reminiscent of the territorial call of an Eastern Screech-Owl.

The food begging calls of the nestlings and fledglings is a thick *pshshshshshsh* which is best described as the sound made when someone opens a soda can. When the owls give this call, it can last a few seconds at a time and of course is given in several intervals.

Courtship

According to McCallum (1994) and Reynolds and Linkhart (1987) courtship begins with the male soliciting the female, away from the nest. The male approaches his potential mate silently, or occasionally by giving a two-note call. He perches close to her and delivers food bill to bill.

Copulation often occurs after a food transfer, as it often does with Northern Pygmy-Owls. The pair can occasionally be seen perched together preening one another. The male feeds the female prior to, and after, egg laying, but the rate of feeding will increase and peak four days prior to egg laying.

Egg Laying and Incubation

The eggs of Flammulated Owls are slightly oval shaped and glossy white. Of 38 eggs, the average measurement was 29.1 to 25.5 mm (Bent 1938). With limited information on clutch sizes available, it appears that these owls lay from two to four eggs. The nesting pairs that I've watched all fledged two and three young respectfully.

Incubation is by the female exclusively and begins after the second egg is laid. It lasts 20 to 24 days and after the eggs hatch the female broods the young exclusively, while the male provides for the family.

Throughout the owls' range, egg-laying and incubation appears to begin between April and July.

The Young

I've been told by other researchers that a day or so before the eggs hatch, you can hear the young chirping from within the egg. Unfortunately, I haven't been in a situation to hear this myself, nor have I come across anyone with documentation that disproves this with Flammulated Owls.

At hatching, the owlets emerge from the egg with eyes closed and their heads and bodies covered with white down. Their feet and bill are flesh colored (Bent 1938). Over the first 10 days, the white natal down

Fledgling Flammulated Owl the first evening out of its nest.

begins being replaced with the birds' juvenile plumage. Then over the next few weeks, the young grow their contour and flight feathers, which they will wear until they molt the following summer.

A few days before fledging, the female leaves the owlets in the nest alone. Around 8:30 each evening, you can often see the little ones looking for their parents to bring them food. During this time, the owlets can often be seen fighting for position at their front door.

While watching the owlets on 19 July 2004, a funny thing happened. As we got to the nest area, I saw the male perched in a live spruce just across from the nest entrance. He was well concealed as he perched up against the trunk of the tree. So, I decided to set up my camping stool 30 feet (9.15 m) or so from the nest. I figured the owlets were close to leaving home for the first time because that was the first time the male was perched so close to the nest.

By 8:28 p.m. both young were peering from the nest entrance as the adults were feeding them. It was interesting to watch the nestlings as they stretched, partially extending themselves out of the nest to get a

After the first owlet fledged, its two siblings wanted to see where it had gone.

morsel of food. Then, at 8:41, one of the owlets overextended itself right out of the nest. It appeared to be pushed from behind as it reached for food. It came completely out of the nest, and then quickly tried to reenter but fell to the ground.

I ran over to the nest tree to see if I could find the little owl and also make sure the little bird was not hurt, but I couldn't find it. As I got close to the area where I thought the little owl was, its mother began attacking me. I quickly moved back out of sight, as the parents continued feeding the nestling.

The following evening, we arrived near the nest and after a bit of searching I located one adult perched up against the trunk of a live spruce several feet west of the nest. I figured it was close to the owlet that fledged the previous evening.

A short time after nine, something caught my eye. As I turned around, I saw the fledgling perched on a branch of a dead spruce 12 feet (3.65m) up.

Knowing that the first owlet had fledged the previous evening, I had brought my bands with me, just in case I was able catch one or both of the owlets. But unfortunately, the little owl was just out of my reach. I wanted to band the little guy in the hopes of gaining some insight into the birds' lives and migration patterns.

All this time, the remaining nestling was continuously food-begging from it's home, so I asked Gary, my research partner, to keep an eye on it, while I tried to get hold of the fledgling. I remembered that, since 1998, there was a small metal bucket under a spruce tree. I retrieved the bucket and placed it upside-down on the ground under the owlet so I could reach it.

I stood on the bucket, which was fortunately just the amount of extra height I needed to reach the owlet. I slowly grasped the bird, then knelt on the ground and proceeded to band and measure it. I then checked the condition of the little fellow. It was quite fat and in great condition.

{Over the years, I've banded several owls that had recently fledged. And they all act the same. They don't seem to have a care in the world. They don't get the slightest bit afraid or nervous.}

We took several photos of the little owl and placed it back on the branch I took it from. Interestingly enough, the adults did nothing during this banding period. We left the area a short time later, and I returned the following evening to find the second nestling still had not taken the plunge. Yet two days after that, there was no sign of any Flammulated Owls.

In 2008, the adult Flammulated Owls raised three owlets, with the first fledging on 20 July. The second fledged on 21 July and the third left home on 23 July. All three owlets were found within 40 feet (12.19 m) of the nest tree the day after they fledged. The first owlet was about six feet (1.82m) from the ground perched against the trunk of a live spruce. The second owlet was perched against the trunk of a Ponderosa Pine about 4 feet (1.21 m) from the ground. The third owlet was located about 3 inches (7.62 cm) from the ground perched on a live spruce branch.

That year I was fortunate to be at the nest the day after the birds fledged.

Description of the Fledglings

At fledging the owlets' juvenile plumage is light gray with slightly darker, fine gray horizontal barring covering the head, beast, belly, and back. You may or may not see a few fine horizontal black bars on the breasts.

The ear tufts are quite small, yet the feathers around the eyes that comprise the facial disk, are present, but not completely developed. The dark half moons that comprise the outer edges of the facial disk are readily identified, giving the birds their owl-like appearance even at this young age. The rictal bristles (stiff feathers around the bill) are quite pronounced and the bill seems to be too big for the bird's face. The primaries and secondary flight feathers as well as the tail feathers (rectrices) are about half grown.

Interestingly enough, at fledging, the owlets, which are smaller than their parents, most often weigh as much and, at times, outweigh the adult birds because the adults are spending so much time feeding the young and less time feeding themselves. In 2008, I was able to trap and band three fledglings, which weighed 61.2 grams, 63.2 grams, and 52.7 grams respectively.

By the second week, the young owls' facial disk, complete with the dark outer edging, are much more prominent, but not fully developed. The ear tufts can be seen protruding above the eyes. Black vertical streaking appears on both the right and left sides of the breast.

The contour feathers on the forehead crown and back begin showing adult-like markings, which means that the shafts of the feathers become black with dark finger-like horizontal bars. These feathers around the bills grow and cover the nostrils. The wing and tail feathers continue growing, as does the definition on the scapular feathers, which gives the birds their name.

By the third week, the owls' ear tufts are well defined and almost completely developed, as are the facial disks. The dark vertical marks on the forehead, crown, back, and belly continue growing thicker. The facial disk begins showing some rufous color around eyes and bill, as well as some fine rufous barring on the breast and belly. The rufous around the scapular feathers is much more pronounced and the wings can be full length by this time.

By the end of week four, the birds look just like the adults at first glance, but the primary feathers are slightly pointed versus rounded as they are with the adult birds. When perched, the primaries will extend past the tail. The facial disk and ear tufts are adult-like and the vertical barring on the body is thick, as is the horizontal barring. Also, by this time, I believe the young are almost, if not completely, independent of their parents as the birds begin their southern migration.

Another fledgling from that same 2008 nest.

The painting depicts an owlet just a few days after it had fledged the nest (left), and the same bird one week later (right). Not only does the bird's plumage change, but its personality does too.

The same owlet three weeks after it has left the nest.

Feeding the Fledgling

After the first owlet had fledged in 2004, I watched the adults feed their offspring several times. The owlet was often perched conspicuously on an exposed limb as the adults approached with food.

The owlet was perched on a limb of dead spruce when one of the adults approached with a moth in its mouth. The owlet walked to the highest point of the branch it was perched on and leaned toward its parent. At the same time the adult came in with outstretched legs and seeming to land on the same branch momentarily.

Interestingly enough, as their bills met, both birds closed their nictitating membranes just long enough to transfer the food from adult to young. The nictitating membrane is often called the third eyelid and is actually a membrane under the bird's eyelids that aid in moistening the eye and protecting it from potential injury, specifically when the bird is attacking prey or feeding young.

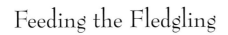

In 2004, while watching one of the fledglings, an adult came in to feed it. The owlet walked to the tip of the branch as the adult swooped in from underneath. When the adult feeds the fledgling, it often closes its eyes and uses the rictal bristled around its mouth to feel where the baby's mouth is.

Migration

Being insectivorous means that in order to find food, Flammulated Owls need to migrate south for the winter, with most researchers agreeing that they winter in Mexico and farther south. However, during their fall migration, migrating Flammulated Owls occasionally shoot a little too far east, ending up in places like Louisiana, Alabama, and even Florida, before moving farther south.

In Colorado, Flammulated Owl migration appears to begin in late August or early September, with a few individuals remaining into October. I would think that a similar timing of migration occurs throughout the rest of the owl's northern range as well.

One Flammulated Owl was captured in Arizona during a snow storm on 31 October 1972, which means that birds occasionally remain in the southern parts of North America through October (Balda et al. 1975).

A Flammulated Owl migration study in New Mexico (Delong 2000) found that the majority of the migrating birds were trapped during the first two weeks of September, but birds continued to be captured through mid-October.

As with fall migration, these birds must move north in the spring to reach their breeding grounds. In New Mexico, banders began trapping migrating birds in April (Balda et al 1975) as they made their way north. Marshall (1939) found birds in California vocalizing in late May, meaning they must have returned to their breeding grounds prior to that. Marshall (1953) located a Flammulated Owl in the Catalina Mountains of Arizona on 26 March. In Nevada they have been recorded from early to mid-May and in Utah, in late April (Balda et al. 1975).

As the owls move north in the spring, they appear to migrate through lower elevations than those in which they will eventually nest due to the lack of insects as a food supply in the higher elevation conifer forests at the time. This way, by the time the owls arrive on their breeding grounds, there is an ample supply of insects available to feed the birds (Balda et al. 1975).

Flammulated Owls, like other migrating birds, appear to have a rapid northerly migration in the spring, yet a more leisurely southern movement in the fall (Balda et al. 1975).

Chapter Four

Rehabilitation of Flammulated Owls

In addition to the Northern Pygmy-Owls discussed in part one, I've had the opportunity to rehabilitate a couple of Flammulated Owls also. My first injured Flammulated Owl was found in the parking lot of the RMNP headquarters in September 1999.

My phone rang about 7:00 a.m., and the person on the other end informed me that she had a small owl that was injured. I asked if they knew what kind it was, which she replied, "I think it is a screech-owl". Knowing that we have no screech-owls at this elevation, I asked if the bird had dark eyes; and she said she wasn't sure because its head didn't look quite right.

I arrived at the headquarters a few minutes later, walked into the book store, and asked about the owl. I was directed to a small room with two park service employees and a small box on a table. After a few formalities, I opened the box to find a Flammulated Owl with a severe eye injury.

The bird's right eye, including the socket, had been almost entirely removed from its head. What remained was a dried eye and socket only connected to the bird by a small (just a few millimeters) piece of skin, holding the eye to the bird's head.

I took the owl home and fed it a few pieces of cut-up mouse then placed it in a small pet carrier and left the bird alone. Later that evening, I went in to check on the little owl to find that it had expired.

Even though wild Flammulated Owls feed on insects, they will eat small bits of meat offered to them. In fact, they will readily consume bits of mice in captivity. Mice are an excellent source of calcium and protein for owls including Flammulated Owls.

Apparently that owl was so emaciated that its intestines had fused together. If the bird eats solid food at that point, the food can actually rip a hole in the intestines and kill the bird. This is why it is important to give fluids to an emaciated bird for the first few hours, so the liquid can open the intestines before the bird consumes solid food.

A few years later the second Flammulated Owl came to me via one of the local veterinarians. I received a call from the vet about a small owl that was brought into the clinic. I couldn't get to the bird immediately, so I asked the doctor to place the bird in a small pet carrier in a quiet area; and I would be there as soon as I could.

I arrived at the vet's office about an hour later as the doctor came out and showed me the X-ray revealing a break very near the bird's elbow. I looked at the owl and instantly identified it as a Flammulated Owl.

I asked for the contact information of the people that found the bird, so they could tell me what happened to the little owl. I'm always intrigued by how the birds get injured.

I took the bird home and set up a pet carrier, which entails placing a dark towel on the bottom of the carrier, along with a small dish of water. I didn't give him a perch for several days. I wanted him to stand directly on the towel so that the wing would remain virtually immobile and he would not injure it further. Further injury might occur if the bird would move up onto and down from a perch.

More often than not, if a small owl comes in with a broken wing, I will wrap the wing with a self adhesive wrap, which immobilizes the wing, allowing the bone to heal properly.

However, due to the type of injury, I purposely didn't wrap the wing because the bird was holding the wing normally. With this type of break, if you wrap the wing, you run the risk of the joint calcifying shut. If that happens, the bird will never fly again.

A few hours later, I contacted the individual who found the owl and she informed me that the bird that I had, along with two others, were roosting in their small shed. When she walked inside the shed, the owls exited and one flew into her Saint Bernard. The owl ended up on the ground as the other two birds were flying around the dog, apparently trying to distract it. The lady picked up the little owl and brought it to the animal hospital.

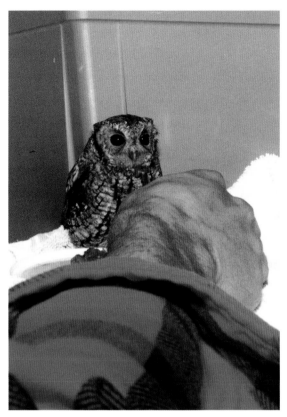

Scott hand feeding an injured Flammulated Owl.
© Susan Rashid

When injured birds need rehabilitation, I try to decorate their hospital cages so that the bird feels more comfortable. At times, I feel this can help aid the bird's rehabilitation.

Due to the Flammulated Owl's nocturnal lifestyle, I decided not to feed him until about 7:00 p.m. At that time, I placed a cut up mouse in the cage with the owl. But, at 9:00, I noticed that the owl hadn't yet eaten. I was getting quite nervous, because I didn't want the little guy to starve.

I checked the bird every 15 minutes or so until 11:30, at which time, I decided to take matters into my own hands. By that time, the food I had placed in with the owl was getting dry, so I took a small mouse from my mouse house, dispatched it, skinned it and cut it into small pieces.

Using a forceps, I grasped a small piece of meat, opened the cage door and slowly moved the meat near the bird's mouth. To my amazement, it instantly began eating. I was quite impressed with myself, being able to hand feed a wild owl. I reenacted this routine every evening for 16 nights. On the 17th evening, I opened the cage door to feed the owl and it flew right past me and landed behind me.

I knew the bird was healed, but by this time it was the middle of October. I figured it would be better for the owl if it were to spend the winter at the Birds of Prey Foundation, along with the resident Flammulated Owl, until the following spring. By then it could be released at a time when the owl wouldn't have to deal with the potential hazards of migration.

The following June, the owl that I had previously named Sprinkles was released, along with a second Flammulated Owl named Hercules. Both were released just at the edge of RMNP about 800 yards (244 m) from the tree that a pair of Flammulated Owls used for nesting in 2004.

On 8 August 2005, I received a call from the owner of a local campground about a small owl that was being attacked by Black-billed Magpies. I asked her to place a small box over the owl and I would be there as soon as possible.

On the way to the owl, I kept thinking that it was going to be a Northern Pygmy-Owl because I had seen them in that vicinity in the past. Also, being 3:00 p.m., it would make sense for it to be a diurnal owl. My wife, Susan, set up a small pet carrier, complete with a towel, and we were on our way. We got to the campground and met the owners who told me the bird had gotten out from under the box and moved off.

I was quite upset because a similar thing happened several years ago after I got a call about a Great Horned Owl that was hit by a car. I told the people to place a towel over the bird so it wouldn't get away before I got there. The people were too afraid of the large owl to get close to it. That owl couldn't fly very well, and as it tried to move, it was hit by a car and killed.

Susan and I moved toward the box and found a little Flammulated Owl a few feet away perched on a brick. I slowly picked it up and saw that it had a few feathers missing from its forehead. It had apparently been hit in the head by one of the larger birds.

I presumed the owl had a severe headache, so I brought it home, giving it a small dish of water, and placed it in a dark room to allow it to rest. I checked on the bird a few hours later and unfortunately it had expired. Its injuries must have been more severe than I thought.

Due to this species' migration habits, when over-wintering rehabilitated Flammulated Owls, we have found that placing a heat lamp over the bird's food and water dishes keep them from freezing while they are outdoors. The birds remain outdoors throughout the winter unless the weather gets below zero, at which time they are brought indoors until the weather warms.

Mortality and Longevity

Before birds of prey were federally protected, Flammulated Owls, along with other owls, were routinely being dispatched for private collections. Spring snowstorms are also apparent causes for a percentage of dead owls due to starvation (Ligon 1968).

An adult Flammulated Owl was found dead on 8 September 1945. It had apparently swallowed a rather large grasshopper the wrong way and ended up suffocating (Kenyon 1946).

Birds of prey, such as Great Horned Owls, Spotted Owls, Cooper's and Sharp-shinned Hawks, as well as house cats have been documented killing and eating these little owls (Richmond et al. 1980, Johnson and Russell 1962, Borell 1937, McCallum 1994).

Some have been caught as late migrants and taken to rehabilitation centers where some have survived to be released the following year, yet others were too emaciated and perished.

A portion of an adult Flammulated Owl was found under its nest after researchers had enlarged the entrance (but replaced the piece they cut) to band the adult and young. After that, an unidentified mammal apparently dislodged the cut out piece and killed the adult (Richmond et.al.1980).

In 2005, I was told about a dead bird on a soccer field of a high school. When I got to the bird it turned out to be a recently fledged Flammulated Owl, but it was so badly decomposed that I could not tell the cause of death. Also, since 2004, Flammulated Owls have perished from West Nile virus.

According to Linkhart and Renyolds (2004), the oldest Flammulated Owl on record as of 2004 was at a minimum 14 years old. Their research has found other individuals that have been recaptured at 11, 9, and 8 years old respectively. Talking with Ms. Sigrid Ueblacker, of the Birds of Prey Rehabilitation Foundation, she revealed she has a captive Flammulated Owl that as of 2008 was 22 years old.

Occasionally these owls crash into windows and perish.

An adult Flammulated Owl (in the front) and a Northern Saw-whet Owl comfortably perching together at the Birds of Prey Foundation in Colorado. These birds often perch close together. The Flammulated Owl in this photo was 22 years old as of 2008.

Scott Rashid

Part Three:

The Northern Saw-whet owl

Aegolius acadicus

The Northern Saw-whet Owl

Chapter One

Discovering the
Northern Saw-whet Owl

The Northern Saw-whet Owl is the smallest nesting owl east of the Rockies and one of the smallest in the West. Throughout their western range, these adorable little owls are often found in the same habitat as the Northern Pygmy-Owls, Flammulated Owls, and occasionally Boreal Owls.

Like Flammulated Owls, Northern Saw-whet Owls are often located in the spring after they begin their nocturnal vocalizing. Unlike Flammulated Owls, Northern Saw-whet Owls will occasionally call during the day, at which time they can easily be observed. Before I began studying Northern Saw-whet Owls, I had always assumed that the species was completely nocturnal, remaining immobile and quiet during daylight. However, certain males will on occasion vocalize during the day, especially when it is overcast.

I have located Northern Saw-whet Owls every year since I began researching them in 1997. By far, the majority of the owls are found or at least heard in the spring while they're vocalizing to attract a mate. In fact, they seem to be relatively common throughout the Estes Valley and Rocky Mountain National Park (RMNP) and I believe that they're the third most numerous owl in the area, behind the Great Horned Owl and Northern Pygmy-Owl.

Interestingly enough, in 2007, while searching in and around RMNP, I was able to locate 10 different calling males and two nest sites during the month of March alone, with two more calling males and a third nest located the following month. Northern Saw-whet Owls begin their nesting season earlier each spring than both the Flammulated Owl and Northern Pygmy-Owls do, but roughly the same time as the Boreal Owls do.

The first nesting Northern Saw-whet Owl that I had come across was located, almost by mistake. It was in the summer of 1998, when I was volunteering for the resource management department of RMNP, searching for small owls within the park. One of the park biologists asked me to take a look at a nest cavity in the Upper Beaver Meadows area of the national park. The cavity in question was in an aspen, which he thought might have been occupied by a Northern Pygmy-Owl.

We had agreed to meet in the Upper Beaver Meadows parking lot early one afternoon and after a moderate hike, arrived at the tree in question. He pointed to a knot in the tree about 24 feet (7.31 m) from the ground that had a few feathers seemingly wedged in it. He thought that maybe a Northern Pygmy-Owl had stored a bird there.

Unbeknownst to him, just a few inches below the feathers was a cavity. After pointing out the cavity to him, I whistled a few toots of the Northern Pygmy-Owl's territorial call, hoping to have located an active Northern Pygmy-Owl nest. But within seconds, a Northern Saw-whet Owl was at the cavity entrance.

The owl perched at its front door momentarily, as I took a few photos. After a few minutes, the owl lowered back into its home as we moved off and hiked back to the parking lot. My intention was to return after dark to begin researching the nesting Northern Saw-whet Owl. Unfortunately, I was only able to return to the nest once before the young had fledged that year.

During their nesting season, Northern Saw-whet Owls are virtually inactive during the day, as the female remains in the nest and the male roosts quietly nearby in dense foliage. Then around 9:00 p.m. each evening, the male begins his nightly forays searching his surroundings for small mammals or other tasty morsels that he will deliver to the female who in turn feeds the nestlings.

Interestingly enough, in 2006 I found a female Northern Saw-whet Owl nesting in a Ponderosa Pine 42 feet (12.80m) from the nest tree that I just described, which told me that Northern Saw-whet Owls find an area acceptable for nesting but not necessarily an individual nest cavity itself.

As they do with all owls, songbirds find Northern Saw-whet Owls a threat to themselves and their young. So when found during daylight, Northern Saw-whet Owls are often harassed by any and all songbirds within the immediate vicinity.

I witnessed this first hand several years ago when I lived near a shallow creek surrounded by Mountain Alder, Quaking Aspen, and Ponderosa Pines. One evening, just before dark, while walking the ditch, I heard some songbirds scolding something ahead of me. As I approached the birds, I noticed a Northern Saw-whet Owl perched on a branch about five feet (1.52m) from the ground. The owl was surrounded by harassing Mountain Chickadees and Pine Siskins.

The particular moment I'll never forget was when a female Broad-tailed Hummingbird hovered just a few inches from the owl's face, yet the tiny predator did not seem to care at all.

I ran the few yards to my house and got my camera, complete with flash, and returned to the area where the owl was last seen. To my amazement, the bird hadn't moved an inch. I was able to take several photos before the bird flew a short distance up the creek.

Having read that these owls can be closely approached and often grasped while perching, I attempted this to no avail. Just as it was too dark to see the owl clearly, the songbirds moved off for the evening and I figured I had bothered the little owl long enough, and moved away.

What's In a Name?

Johann Gmelin, a European explorer, discovered the first Northern Saw-whet Owl in 1788 in what is now Nova Scotia. The birds' scientific name, *Aegolius acadicus*, is derived from *aegolius*, meaning a kind of owl, and *acadicus*, from the area of the country where the bird was first found.

Northern Saw-whet Owl hiding in a Ponderosa Pine at the Birds of Prey Foundation.

Distribution and Range

The Northern Saw-whet Owls' breeding range is restricted to North America, with breeding documented in the mountains of eastern Tennessee, western North Carolina, and the mountains of West Virginia. Then there appears to be a gap in breeding until northern Illinois, Indiana, Ohio, and southeastern New York State through central New York. From Connecticut north through Maine, Nova Scotia, northern New Brunswick, Southern Quebec, central Ontario, central Manitoba, central Saskatchewan, and central Alberta, then north through central British Columbia to southern Alaska.

Throughout the western United States, Northern Saw-whet Owls range from western Oregon, through most of California, to a portion of northeastern Nevada, eastern Utah, Arizona, western Colorado, and western New Mexico (Johnsgard 2002).

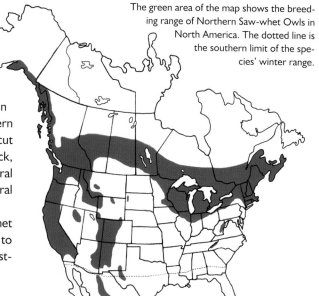

The green area of the map shows the breeding range of Northern Saw-whet Owls in North America. The dotted line is the southern limit of the species' winter range.

Chapter Two

Anatomy of the
Northern Saw-whet Owl

I generally reserve the word cute for things like kittens or puppies. However, cute definitely defines this little owl.

An adult Northern Saw-whet Owl has a comparatively large, rounded head, without ear tufts and large yellow eyes, which in some older individuals can be golden. It has a black bill with white "V" shaped feathers above, which extend between and above their eyes. The Northern Saw-whet Owls' facial disk is brown with white spoke-like feathers extending outward to the edges of the disk. Just behind the facial disk lie the ear openings, which are remarkably asymmetrical, the right ear being larger and higher than the left. Northern Saw-whet Owls like other owls have moveable earflaps that can cover the ear openings when needed.

An Adult Northern Saw-whet Owl.

Looking at this bird's eyes, you can see it's nictitating membrane. This membrane is often called a third eyelid; it helps lubricate and protect the eyes.

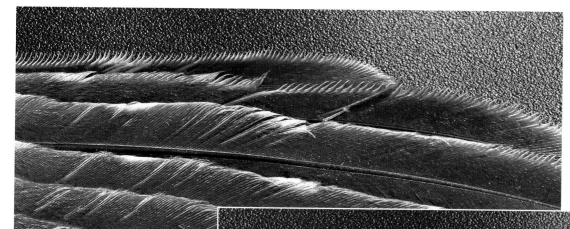

Close-up of the fringed edge of a Northern Saw-whet Owl's primary flight feathers.

Close-up of the edge of a Black-billed Magpie's flight feathers.

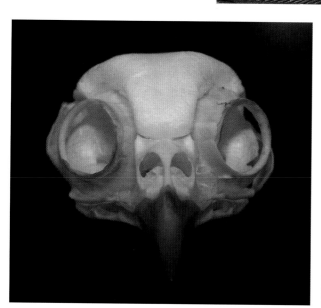

The skull of a Northern Saw-whet Owl. Note the uneven or asymmetrical ear openings. The birds' right ear is just behind its eye and the left is below the other eye.

Close-up of the ear opening of a Northern Saw-whet Owl.

The forehead is brown with small vertical white feathers which look like little streaks of paint. The nape of the neck has large white spots somewhat resembling the false eyes of a pygmy-owl, yet, in Northern Saw-whet Owls are not as pronounced.

The breast and belly is white, with reddish-brown vertical streaks. Their feet are feathered to the talons, which are black.

With the exception of their breasts, bellies, and faces, the overall color of the birds is a rusty-brown. The top side of the wings, back, and tail are a kind of raw umber or reddish brown, with a few to several white spots. The white spotting is quite evident on the bird's scapular feathers. Also, the tail is short with white horizontal barring.

They have long rounded wings with 10 primary flight feathers, which extend slightly past their tails when perched. The leading and trailing edges of the flight feathers are fringed like a comb allowing air to flow through the edges of the wings and muffle the flight, making the owls flight virtually inaudible.

Northern Saw-whet Owls are a bit larger than both the Flammulated Owls and pygmy-owls, with most literature noting them anywhere from six-and-a-half to eight inches (16.51 to 20.51 cm) from head to tail and a wingspan of about 17 to 22 inches (43.18 to 55.88 cm). Males are generally smaller than females, but there is some size overlap within the species.

When trying to conceal themselves, Northern Saw-whet Owls often raise the upper corners of their facial disk in such a way that the top of the head is in the shape of a shallow "M". They will also, occasionally, pull a wing over their breasts and bellies as well.

Vocalization

The most frequently heard call given by Northern Saw-whet Owls is a single note whistle-like *toot* or *whoop* or even *kwook* (Bent 1938), which on occasion can sound very similar to that of the Northern Pygmy-Owl. However, there are a few ways to discern between the calls of the two species.

The call of the Northern Saw-whet Owl is an ongoing *toot* which can go on for an hour or more without stopping. It is often described as

A Northern Saw-whet Owl in
its concealment posture.

Northern Saw-whet Owl in typical calling posture. © Bill Schmoker

giving three-notes-per-two   seconds, whereas the call of the Northern Pygmy-Owl (depending on race) is roughly one *toot* per second and seldom continues for more than five minutes without stopping. On the other hand, I timed one individual Northern Saw-whet Owl that vocalized continuously for 43 minutes before stopping.

Another difference between the two species is that the Northern Saw-whet Owls most often, but not exclusively, begin calling each evening after dark when the Northern Pygmy-Owls are no longer active. On several occasions throughout the years while searching for Northern Pygmy-Owls, I've heard several Northern Saw-whet Owls vocalizing during the day. I believe that the calling owls were young birds that were trying to entice a mate for the first time. In 2006, there was a little male that would call both day and night from within the concealment of a dense stand of Ponderosa Pine trees on the edge of town. After a little more than two weeks the owl was no longer heard. Presumably, it was unable to attract a mate and moved to another area in hopes of increasing its odds of attracting a mate.

William Brewster (1925) describes a few of the Northern Saw-whet Owls vocalizations as follows, "On May 18th, they were given at infrequent intervals and always in sets of three thus: *skrigh-aw, skrigh-aw, skrigh-aw.* Their general resemblance to the sounds produced by filing a large mill-saw was very close, I thought."

On May 28, he heard a somewhat different, metallic note. The owl kept it up a little more than a minute, regularly uttering four apparent monosyllabic notes every five seconds. "Their metallic quality was so pronounced and their tone so ringing that they reminded me of the anvil-like *tang-tang-tang-ing.*" Another described vocalization is a kind of *whurdle-whurdle-whurdle,* which is similar to a call made by a pygmy-owl but a bit more guttural (Bent 1938).

In most cases, if you hear a calling Northern Saw-whet Owl, you're often within about 300 yards (273 m) of the bird. This is because Northern Saw-whet Owls often vocalizes from the concealment of conifers such as pine or spruce, which absorb the sound waves and protects the small owls from the prying eyes and talons of larger owls such as the Barred Owls, Great Horned Owls or Spotted Owls. When Northern Saw-whet Owls vocalize, they often lean slightly forward and toot, at which time their throat inflates with each toot. Occasionally, they seem to inflate and deflate with each toot.

Conversely, the call of the Northern Pygmy-Owl often carries a half mile or more on a windless afternoon. Pygmy-Owls, unlike Northern Saw-whet Owls, often vocalize on an exposed limb, sometimes as high as 60 feet (18.29 m) from the ground. With no obstacles to absorb the sound waves emitted by the pygmy-owl, its call will ultimately carry a significantly farther distance.

However, in certain situations, the call of the little Northern Saw-whet Owl can carry quite far. On 18 February 2006, at 8:45 p.m., while on our front porch, my wife and I heard a calling Northern Saw-whet Owl south of our house. As I tracked down the owl, I found it almost three-quarters of a mile from our house. It was perched on a limb roughly 50 feet (15.24 m) from the ground within a small clump of Ponderosa Pine trees. The habitat between the calling bird and my yard was primarily open horse pasture with few houses, so there was nothing to absorb the sound waves.

Another interesting characteristic of the Northern Saw-whet Owls' voice is that it often has a ventriloquial quality, making the bird appear to be in a different place than it actually is. I noticed this for the first time in the spring of 2004. While sitting on our front porch my wife and I heard a Northern Saw-whet Owl calling from what seemed to be just a few yards east of the house. So, I walked the couple of hundred yards down the street toward the calling bird. But, when I got to the corner of the block, I couldn't hear it anymore. Therefore, I assumed that the bird had just stopped calling and I walking back to my house. But before I got half way home, I was hearing the bird again.

At first I figured that maybe the bird stopped vocalizing before I got to it. Then I decided that the cliff at the end of the road might have been deflecting the owl's voice in such a way that I couldn't hear it from the bottom of my street.

The next evening, the owl began calling just after dark. This time, I walked down the hill around the cliff to find the bird calling several hundred yards away. After a bit of nighttime hiking, I located the owl calling from a nest cavity in a Ponderosa Pine tree.

This ventriloquial quality, at least in that instance, occurred because the pine trees were scattered throughout the property, causing the sound to bounce off the adjacent trees. Once I got onto the property, it took about seven minutes before I could pinpoint the calling bird. Like the Northern Pygmy-Owl, Northern Saw-whet Owls often solicit a female while inside a cavity.

What Time of Year Do Northern Saw-whet Owls Begin Vocalizing?

Knowing that Northern Saw-whet Owls breed locally in Colorado in November in 2004, I made a conscious effort to determine when they begin their spring courtship vocalizing.

What I found astounded me. Through an Internet listserve that I belong to, I'm in contact with several birders throughout Colorado, one of whom e-mailed and mentioned that he had a Northern Saw-whet Owl calling at about 6:00 p.m. in November and again during the week of Christmas 2005. However, that may have been a young bird of that year which was just practicing his courtship calling before the real courtship began.

Later that same winter, there was another bird, presumably a male, calling in mid-January near Drake, Colorado, at roughly 6500 feet (1982 m) above sea level. Two weeks later, near Estes Park, Colorado, about 15 miles (24.19 km) up the road and 1000 feet (3280 m) higher in elevation, another bird – again presumably a male – had begun vocalizing.

In and around RMNP, the majority of Northern Saw-whet Owls begin a normal routine of vocalizing in about mid- to late February. Yet by the third week in May, most, but not all the owls seem to cease any vocal activity.

In March 2004, while searching for Northern Pygmy-Owls within RMNP, I heard a Northern Saw-whet Owl calling early one evening well before dark. It was just a few yards from a previously used Northern Pygmy-Owl and Flammulated Owl nest tree. Unfortunately, the bird didn't nest in that area that year. However, three years later, a pair of Northern Saw-whet Owls nested in that tree with incubation beginning the last week in March. This meant that these three species prefer the same habitat. To illustrate this a little further, Holt and Leroux (1996) found a Northern Saw-whet Owl and a Northern Pygmy-Owl nesting in the same tree at the same time and the young of both species fledged.

While, searching the Cow Creek area of RMNP in March of 2005, I heard, at 6:25 p.m., a Northern Saw-whet Owl calling from the east side of the gravel road just past the park entrance.

Northern Saw-whet habitat within RMNP in February.

It was giving its distinctive whistle-like *whoop....whoop.....whoop*, which seemed remarkably loud for such a small bird. Yet, occasionally that call was so subdued and soft it was almost inaudible. It would intermittently separate its *toots* with a short pause, presumably listening for other calling owls. When calling after dark, the bird often perches between six and 15 feet (1.83 m to 4.57 m) from the ground and is either perched near the trunk of the tree or near its edge in among dense needles.

Throughout the evening, the bird called for several minutes at a time, then moved either north or south along the west facing slope, stopping every few yards to seemingly mark its territory with some sort of verbal sign, telling any males in the area that this portion of the park was taken. Yet, at the same time, he was advertising for any available females to come on over and join him.

Two nights later, I returned to that same area, and to my elation, found two calling Northern Saw-whet Owls in that same general vicinity. I presumed that the calling bird was the same individual that was calling a few evenings earlier and most likely a male. I assumed the second bird was a female because its call was quite a bit softer than the one that I figured to be the male. At one point, both birds were perched together cooing softly and preening each other, which continued for about two minutes before they flew off in opposite directions. I returned to that area several times within the following weeks, but unfortunately was unable to locate a nest.

After hearing this bird calling, I found it just a few feet from what turned out to be its nest in 2005.

Adult Long–eared Owl

A pair of Northern Saw-whet Owls perched side by side.

There are several regions within RMNP that seem to be perfect habitat for these owls. Each year I check these places in hope of finding calling owls. While searching the park in mid-March 2006, I found a calling Northern Saw-whet Owl just west of Little Horseshoe Park. Since the early 1990s I had thought that particular area would be an ideal location for Northern Saw-whet Owls to use for nesting. It has all the components that the birds seem to prefer; yet until 2006 I had not found any owls there. Much to my excitement, as I approached the spruce trees early one evening, a Northern Saw-whet Owl was soliciting.

As I was walking up the road toward the calling owl, I heard the single hoot of a Long-eared Owl several yards east of the smaller Northern Saw-whet Owl. While the Long-eared Owl was vocalizing, a pair of Great Horned Owls began calling even farther east yet. Shortly after the Great Horned Owls began calling, the Long-eared Owl stopped, but the Northern Saw-whet Owl continued.

The Long-eared Owl was heard in that area for the following three nights but wasn't heard after that. It's possible that the Great Horned Owls stopped the smaller owl's vocalizing by consuming it, or the Long-eared Owl simply moved to another area. The Northern Saw-whet Owl continued calling for three weeks, then it was no longer heard or seen again.

Individual Northern Saw-whet Owls have been heard calling for a period of 70-93 days before nesting began, with an average being 81 days (Johnsgard 2002).

In March 2007, while hiking along the Cub Lake trail inside RMNP, some friends and I found a Great Horned Owl calling. We walked toward the calling owl to find it perched atop a Ponderosa Pine. As the owl called, it would lean forward, lift its tail, and hoot. As it calls, you can see its attractive white throat quiver.

While the Great Horned Owl was calling, a Northern Saw-whet Owl began calling just a few hundred yards away. Then, after a few minutes more, a Long-eared Owl began giving its single note hoots. Ironically, all three species were calling within about 200 yards (60.97m) of each other. The habitat that the Great Horned Owl was calling from was open Ponderosa Pine and juniper. Between the Great Horned Owl and the Northern Saw-whet Owl were several yards of Mountain Alder with a stream running through the center. The Northern Saw-whet Owl was in a dense mixed aspen and spruce-fir section on a north-facing slope. Farther up that same slope was the Long-eared Owl.

The following evening, as I returned to the same area, the only owl heard was the Northern Saw-whet Owl.

Other Vocal Descriptions

Northern Saw-whet Owls call with little concern to low temperatures, light to moderate wind, and/or cloud cover. They will even call during moderate snowstorms.

Some researchers suggest that Northern Saw-whet Owls will increase vocalization on full moon evenings. Personally, I haven't found this to be necessarily true. I have been out during full moons and found calling owls, and have been out on other full moon evenings when I couldn't find a calling bird. So, I'm not sure what, if any, difference a full moon may have on the calling owls, at least in the areas I have researched.

After nesting has begun, the males cease their territorial calling. The only calling heard at this time will be when the male is delivering food to the nest. As he approaches, he will utter a short series of soft, rapidly whistled notes termed the visiting call (Johns et al. 1978).

Another call heard occurs when the owls feel threatened. At this time, you can often hear a kind of buzzing call, reminiscent of a grasshopper.

In Colorado, the owls have been found calling from 6000 feet (1830 m) in aspen and Ponderosa Pine habitat to near 10,000 feet (3049 m) in spruce-fir forests. Interestingly enough, I seldom find Northern Saw-whet Owls calling in the same areas more than one year at a time. I have on occasion, heard them calling within the territory of the smaller Northern Pygmy-Owl as well as the larger Boreal Owl.

During the fall and early winter, I trap and band the owls as they move through the area where I live. The traps are mist nets and the lure is a recording of the bird's territorial call. On several occasions, as the tape was playing, the owls within the vicinity of the nets would give a high pitched *mewing* call, which is a kind of agitation vocalization.

Once in hand, the birds, at times, will give a high pitched *trill* or *chitter* call due to their agitation at being in someone's hand and unable to fly away.

Chapter Three

Courtship

Identifying courtship activities of a nocturnal creature can be a bit difficult, but not impossible. Figuring out what Northern Saw-whet Owls are doing during courtship, takes a lot of time and a bit of luck.

Courtship begins with the male calling incessantly. He will often move through his chosen territory vocalizing as he travels from perch to perch. This move can be just a few feet to 50 feet (15.24 m) or more each time. On occasion, he will be calling with a mouse, vole or other item in his grasp just in case a female comes along.

When a female appears, the male often gives a very rapid burst of staccato notes while flying around her. As he approaches his potential girlfriend, she may respond with soft *swee* notes, which often rise in pitch.

When perched together you may hear the bird's duetting, at which time you should notice that her voice is often quite a bit softer than his. If you're close enough to the birds at that time, you'll hopefully hear the male's staccato calling notes that reach a crescendo of *chuck-like* notes (Webb 1982).

On other occasions, the male either calls from his chosen cavity or near it in an attempt to entice his would-be bride to enter the cavity, which will often have a supply of mice or other tasty morsels inside. When the male is perched inside the nest calling for a female and she comes close, he will exit, allowing her to enter and, he hopes, accept.

In 2003, I found, on several occasions, a male calling from his nest cavity. He was heard and seen for seven straight evenings while perched at the nest entrance calling. He began just after dark each evening, which is when I noticed him, so he may have exited the nest and continued calling from somewhere else within his territory later each evening.

In the western mountains, these owls often call high on steep slopes, which I believe to be intentional. The owls are probably perched high above their nest tree, so they can glide down slope to their chosen nest tree or to capture prey. With this downward glide, the birds may actually be conserving energy.

Nesting

Like the other small owls Northern Saw-whets are secondary cavity nesters seeming to prefer cavities excavated by Northern Flickers. However, they have used Pileated Woodpecker and Hairy Woodpecker cavities, as well as nest boxes and natural cavities. In Colorado, I have found Northern Saw-whet Owls nesting in cavities within aspen, Ponderosa Pine, and nest boxes that had been erected specifically for them.

The first Northern Saw-whet Owl nests that I worked with were near the Dinosaur National Monument in northwestern Colorado. The birds were nesting in nest boxes that had been erected several years earlier. I went with Dr. Ron Ryder specifically to band the nestlings.

The previous year, those boxes were used by Pack Rats which constructed their nests using cactus. It was a surprise to find the little owlets surrounded by cactus thorns. Yet the birds did not seem to mind in the slightest.

I remember opening one nest box to find seven owlets of various ages inside. The oldest was about to fledge while the youngest still had its eyes closed and was unable to stand. All the boxes had a supply of mice within, many in various stages of decomposition.

When we arrived at the last nest box, we found the female inside with her family. While I was climbing the ladder she looked out at me then quickly exited before I reached the box. We banded the six nestlings and also found five Deer Mice inside the nest box.

We decided to return the following morning for a second attempt to catch and band the adult. Again, we missed, but I looked inside the box to see what the adults brought for the family that evening. I found eight mice this time. Figuring each nestling ate at least half a mouse that night, it suggested to me that the male is an extraordinary provider.

Northern Saw-whet Owl looking from its nest cavity inside Rocky Mountain National Park. © Wayne Johnston

This owl was nesting within Rocky Mountain National Park, in the same tree that both a pair of Flammulated Owls and Northern Pygmy-Owls had nested in a few years earlier.

This Northern Saw-whet Owl is peering from a nest within Rocky Mountain National Park.

The third week of May 2005, I was told about a nesting Northern Saw-whet Owl within RMNP. As with the others, it was in an aspen about 21 feet (6.40 m) from the ground in a cavity excavated by a Northern Flicker. This nest, as with the others, was just a few feet from a running stream. It was in an aspen grove that is checked yearly as part of an aspen cavity project. However, that was the first time any small owl was identified nesting there.

The nest tree was in a grove of aspens that stretches almost a half mile east to west. Some of the trees are 40 feet (12.20 m) or higher and virtually all them have the black warty marks that remain after the elk have eaten the bark of the trees. The nest tree itself is on the center of the forest edge. That female Northern Saw-whet Owl began incubating during the last week of March 2007.

A Dr. Ralph found several owl nests in New York State in 1886 (Bent 1938). The first was on April 7th in a dead maple stump 22 feet (6.71 m) from the ground. Another was found in a dead stump 40 feet (12.20 m) from the ground on April 21 and it had five young owls in it. A third was found in a dead stump 20 feet (6.10 m) up. A fourth and fifth nest were found as well both in the last week of April. One was 50 feet (12.24 m) up and the other was 63 feet (19.21 m) from the ground both in dead stumps.

Researchers have told me that it's quite important not to flush a nesting Northern Saw-whet Owl from her nest cavity during the day, because there's a good chance that she could exit the nest and not return until dark. If this occurs while the nest has eggs or owlets that are too young to regulate their own body temperature, you run the risk of the eggs and/or young perishing from exposure.

However, I've read a few historical accounts that claim the owls did in fact return to the nest after having been disturbed (Bent 1938), but I wouldn't want to take that chance.

An interesting aspect to the nesting habits of these owls is that males will occasionally raise more than one family at the same time. When a male (of any species) raises two families simultaneously, it is termed polygamy. Incidentally, I know of a male Barn Owl that had raised two families simultaneously. This only occurs when there is an over-abundance of food, enabling the male to feed both families. The female Barn Owls were nesting in a sandstone cliff a few yards apart and a single male was observed delivering food to both females (Ronald Ryder pers. comm.).

Polygamy has been documented in the Northern Saw-whet Owl as well. Marks et al. (1989) found what is believed to be the first record of trigamy in Northern Saw-whet Owls though. They found a single male that was feeding three females in three different nests during the same nesting season.

As a side note, in 1999, I had erected a nest box for Mountain Bluebirds on private property. The landowners called me one morning, because they were seeing some interesting activity. When I arrived at their property, I was watching a pair of Mountain Chickadees feeding four nestlings inside the box. I also watched a male Mountain Bluebird feeding the nestling chickadees. The bluebird was feeding the chicks grasshoppers, while the adult chickadees were feeding their young spiders. The four nestling fledged, and the male bluebird was seen feeding the fledglings, as were the adult chickadees.

Another story from the "weird nest-fellows department:" years ago, Scott Weidensaul, a Northern Saw-whet Owl researcher, photographed a female Northern Cardinal that was feeding a clutch of American Robin young – even taking the food from the (obviously bewildered) parent robins, then transferring it to the robin chicks. Apparently her own nearby nest had been destroyed a few days earlier when a hedge was cut down (Scott Weidensaul per. com.).

Eggs and Incubation

Northern Saw-whet Owl eggs inside a nest box.

Northern Saw-whet Owls lay from one to seven eggs, with the average clutch size being five or six (Johnsgard 2002). In and around RMNP I've found that three seems to be the average. Their eggs are between round and oval in shape and pure white and average 29.9 to 25 millimeters (Bent 1938).

They appear to lay their eggs in two to three day intervals with incubation beginning as the first egg is laid. Incubation is by the female exclusively and appears to be between 27 and 29 days (Cannings 1993). During incubation and brooding of the young, the female keeps the nest quite clean by removing pellets and feces (Cannings 1993). As the young begin to grow, you can often find excess prey items inside their nest cavity, at which time flies appear near the nest entrance.

The eggs are laid, hatch, and the young grow at different stages, which is believed to be a survival tactic. As the chicks hatch on different days, the young subsequently grow at different stages. This way if there is ever a food shortage, the older chicks will get the food and the youngsters will starve, die, and probably be consumed by the older chicks. This way only the strongest owlets will survive to fledge.

Before the eggs are to hatch, soft calling notes can be heard from within the egg itself (Cannings 1987). I would presume that the time of hatching would coincide with the time of egg laying, meaning the first egg laid would be the first one to hatch.

Throughout the owls' range, eggs are laid between March and May. In Colorado, eggs are often laid anywhere from the last week in March through the last week in April. In Wisconsin, eggs are often found from mid- to late April (Follen 1982). As in Wisconsin, eggs are laid in Maryland from mid- to late April (Brinker et al. 1993). In New York and New England, eggs are often laid between the third week in March and early July (Johnsgard 1988). In Montana, egg laying begins in mid-April, but researchers have found owlets fledging in early August, meaning the eggs are laid much later than April.

Chapter Four

My Research Begins

You never know what's going to occur when watching nesting owls after dark. The nocturnal activity phase of Northern Saw-whet Owls makes them quite a bit more difficult to observe than the diurnal Northern Pygmy-Owls. Watching their nesting activity can create some interesting situations, the most important being identifying what the male was bringing to the nesting female.

Since the first nest I found was near a paved road within RMNP, I didn't want to draw attention to myself and the nest. When I was sitting near it, I could watch vehicles moving up and down the road. Not wanting to draw attention to myself, I would only shine my light when there were no vehicles driving past. This way I knew that no one would be able to see my light while I was in the woods. If anyone would see my light in the woods, there would be a good chance that they would notify the park service, which in turn would investigate and mess up my research. So, to avoid that, I reduced my light shining significantly and only used it when I knew no one would see it.

A Northern Saw-whet Owl nest site in Colorado.

Each evening, I parked my car in a small dirt parking lot alongside the road and walked across the field of knee-high grass and over a shallow creek to reach the nest tree. I then perched on a small camping stool near the nest, which allowed me the ability to see my car and the nest simultaneously.

I chose to sit slightly south and west of the nest so I could use the reflective skylight from Estes Park to help view any activity.

The first night I was watching the nest, my car was approached by park ranger. I had already set up near the nest when I noticed a patrol car pulling off the road and parking behind my truck.

Then the officer began shining a light on it, so after a few minutes, I decided to hike across the field back to my truck and explain to the ranger that I was a park researcher and I was researching small owls within the park. I then asked the officer if he could inform the park service dispatch of my truck's description and license plate number so this would not happen again. Unfortunately, the same thing happened the following evening, but was cleared up after that.

Monitoring Nesting
Northern Saw-whet Owls

Monitoring this species was much more difficult than I had expected. Unlike the Flammulated Owl, Northern Saw-whet Owls only return to their nests a few times each evening. The main difference in the species is that mice and voles can be a bit more difficult to locate and capture than the insects that the Flammulated Owl feeds its family.

What this means, is that there were nights when the male Northern Saw-whet Owl would be heard, then seen delivering prey only once per hour. There were other evenings when he was seen just after dark and several times thereafter.

Prior to the commencement of my Northern Saw-whet Owl research in 2006, I placed orange cellophane over the front of my flashlight as I did when watching the Flammulated Owls, so not to disturb the female Northern Saw-whet Owl too much.

As I did with the Flammulated Owls, I'd arrive at the Northern Saw-whet Owl nest about 8:30 p.m. each evening, so I would be seated before any activity began. More often than not, the female would be at the nest entrance when I arrived. She would remain there, looking out, presumably for her mate until almost 9:30, at which time she would exit unless the male brought food before that.

Once the female exited the nest, I would periodically shine my light near the cavity until I saw her at the entrance again. I only remember seeing her actually reenter the nest twice. Both times she would perch at the entrance before entering.

Each time she exited the nest, it was always a very rapid flight, during which time she would almost hit the ground before flying up into the woods. She would remain out of the nest for up to 15 minutes before reentering. On a number of occasions, she would reenter her nest without my seeing her do so.

Most nights, between 9:00 p.m. and 9:30 p.m., the male would give a few soft *toots* before delivering a mouse to his mate. If she had not reentered the nest, he would deliver the morsel to her and she would bring it to the nest. If she had already reentered the nest, he would wait at the nest entrance while she reached to him and grasped the item from his bill.

However, there were some evenings when he would be silent until almost 10:00 p.m. or even later. He seemed to remain quiet during his hunting forays and only call when he was about to deliver a prey item to the awaiting female.

The area immediately around the nest was primarily an aspen forest with downed logs, grass and a running creek that flowed west to east. Straight south of the nest a few yards were Ponderosa Pine, Douglas fir, and aspen with a number of ground juniper and downed logs.

During the day, I had searched the area within 100 yards (30.48 m) of the nest looking for the roosting male. Unfortunately, I was unable to locate his day roost, although other researchers have on occasion located the day roost of the male, which was often in the same spot each day or at least most days and under which regurgitated pellets and white wash are often found.

That nest was discovered on 27 May 2005. However, at the time of discovery, I had no way of knowing how far along the owls had been in their nesting progression. Then after the first owlet left home one early June evening, I did some research on incubation and the amount of time the young remain in the nest, guessing that the first egg may have been laid on or about 10 April, with incubation probably having begun at that time.

The first of June was when I originally began noticing flies and ants at the nest entrance. Like the Northern Pygmy-Owls, the male Northern Saw-whet Owl is such a good provider that he often delivers more food to the family than the young can consume. If the female doesn't remove it, the remaining food will attract flies and ants, making the nest cavity quite easy to locate.

Perched adult owl about to fly off for an evening of hunting. © Steve Morello

The male is the sole provider for the family during the early stages of the nesting cycle. When he has captured something for his family, he will toot from a nearby tree before flying to the nest and delivering the prey to the female, who in turn feeds the young. Each time I was able to watch this delivery the prey item was a Deer Mouse.

In 2007, I was fortunate to borrow a night vision monocular from a friend, making the night observation a cinch. I still arrived at the nest before dark, but I could sit several yards from it and watch using the night vision in such a way that there was very little disturbance to the adult birds. Using this optic, I could see the adults coming and going from the nest and they seemed to pay me little mind. It was interesting seeing how well the birds navigated their territory when, at times, I could see virtually nothing.

What I found using the night vision that I had been unable to see before was the way the male delivered the prey to the female and later how both adults delivered food to the young in the nest. The male transfers food to the female mouth to mouth. He carries mice or other prey items to the nest primarily in his feet, but at times carries them in his mouth. Regardless of the prey item, the transfer to both female and young is mouth to mouth.

The Young

At hatching, the owlets are blind and helpless, completely dependent on their mother for food, warmth, and protection. At this age, the owlets are covered with white down. For the first 10 days or so you can see the egg tooth on the tip each young owl's bill. This egg tooth is used by the owlet to crack open the egg from inside.

The owlets' eyes open when the birds are between seven and 10 days old. During the next 13 days, you can see the pinfeathers and primaries begin to show growth (Cannings 1993). While in the nest, the owlets grow at a steady rate, owing to the food that the male brings to them.

The nesting pair that I monitored in 2006 had its first owlet fledge on the evening of 7 June, after which I was able to locate the fledgling each evening quite easily. All I had to do was to listen for the owlets' food begging calls, then walk to it. The third fledgling took the plunge on 14 June, with the second owlet having fledged sometime between those two dates.

The female owl apparently remains within the nest during the day until the last owlet fledges. Once out of the nest, the owls do not appear to reenter it.

At fledging, the owlets are two-toned, chocolate brown on the head, upper breast, and back. The belly, legs, and under tail coverts are yellow ochre. Between their eyes, the young owls have beautiful white feathers that extend from their bill up over each eye and below the bill outward. These white feathers that extend both above and below the bill form a kind of "X" on the bird's face with its mouth being directly in the center. There are also a few flecks of white on the birds' foreheads extending from the eyebrows.

The facial disk is similar in color to the head, but the outer edge of the disk is defined by a lighter brown tone. Their eyes are a pale yellow, with the bill and talons being black.

When the wings are closed, the primary flight feathers barely extend past the secondaries. There are a few white flecks on the upper wing coverts, but the back of the bird is unspotted.

The owlets seem to retain this juvenile plumage without any substantial color change until approximately the second week in July. At that time they begin the transformation into their adult plumage. This simultaneously occurs as the white spotting on the forehead virtually extends to the top of its head and, possibly, even further. The primary and secondary flight feathers lengthen and the scapular feathers, which were a chocolate-brown, now begin turning a light tan.

The chocolate color of the breast begins to extend in thick vertical streaks down toward the belly, which becomes a few tones darker. The portion of the facial disk closest to the bill is becoming light tan to white. This white extends straight down from the outer edge of the eyes to the bottom edge of the facial disk.

After that, the white spokes that form the facial disk are clearly evolved. The white spotting on the forehead, top of the head and back of the bird are about complete. The breast and belly are a very pale tan to almost white throughout. The burnt umber streaking of the breast and belly extend about half way down the bird. This brown vertical streaking extends from the breast halfway to the belly. The wings and tail are the same length as the adults at this point.

About five weeks or more after the change begins, the owlets are almost adult in coloration. The toes of the young owls can have a few less white feathers than the adults; however, you may notice a bit of tawny color on the bellies, between the birds' legs.

When we (banders) capture first year owls on their fall migration, which is usually from September through November, we can easily tell the young of the year from the adults because the adults have molted some primary and secondary flight feathers, leaving older feathers worn yet new ones fresh. The young of the year have not yet begun molting, leaving all of their feathers the same color brown.

From the time the birds are wearing their first adult wardrobe through the beginning of their first molt is the only time the birds will have a completely symmetrical coat of feathers. After the birds finish their first molt, the feather color of the birds will no longer be uniform.

Northern Saw-whet Owlet looking out of its nest.

These two owlets were inside the same nest. The oldest is about to fledge, yet the youngest cannot even open its eyes or stand.

Nestling Northern Saw-whet Owl just a day or two before fledging (captive situation).

Recently fledgling Northern Saw-whet Owlet perched against a Ponderosa Pine.

A depiction of what an owlet looks like around the last week in July. Note the extensive white on the face.

First year Northern Saw-whet Owl, often called a hatching year bird. This bird was captured during one of my fall banding sessions and placed on that branch.

This painting depicts an owlet during mid- to late August. This owlet has almost completed the transformation into its adult plumage.

Chapter Five

Hunting and Diet of the Northern Saw-whet Owl

Due to the Northern Saw-whet Owls' nocturnal hunting and feeding, identifying the evening activities are often difficult. However, researchers have found that they most often forage along forest edges, creeks, and streams, and occasionally around buildings and farm houses. They seem to hunt in any habitat that allows them to perch near forest openings.

Northern Saw-whet Owls are seldom found active during the day, but I'd assume the birds may hunt on overcast days during winter to meet their nutritional needs. Their primary activity period appears, however, to be from roughly a half-hour after dark through the evening and ends just before sunrise.

This species has a wait and pounce method of hunting. They most often perch on low tree branches, bushes, or fence posts, often less than 10 feet (3.05 m) from the ground. Then as a mouse or vole is located either by sight or by the sounds it emits, the owl makes its move, flying in rapidly and grasping its victim, often biting it in the back of the head, then flying off to consume its prize in the safety of dense foliage.

Hunting/Food Habits

Northern Saw-whet Owls have excellent visual acuity as well, allowing them to see well in low light conditions. However, the birds may not see as well in extremely bright sunlight.

Their primary food source appears to be small animals such as mice, young rats, young squirrels, chipmunks, shrews, and bats. They also prey upon a few bird species and carrion if needed (Bent 1938).

Prey is grasped with the owls' feet and killed with a bite to the back of the head, unless the bird's grip has already dispatched it. Small items such as young mice or shrews are occasionally swallowed whole, where a larger mouse often lasts for two meals.

George Nicholas in 1921 (Bent 1938) wrote of a flying squirrel that was found in the stomach of a dead Northern Saw-whet Owl. As he dissected the owl, he found the squirrel was swallowed whole. A Northern Saw-whet Owl was even seen feeding upon a dead hare one winter (Bent 1938).

A study done during the breeding season in southwest Idaho (Mark et.al.1988) found Northern Saw-whet Owls feeding on shrews, House Mice, Harvest Mice, Deer Mice, and Montane Voles.

During the winter of 1982, the diet of these owls was studied in north-central Washington State (Grove 1985). Prey items found during that study included shrews, various mice species, and Meadow Voles. The list of birds that these owls preyed upon included White-crowned Sparrows, juncos, House Sparrows, and a Northern Pygmy-Owl.

In 1993 a study of nesting owls in Maryland found Swamp Sparrows, Woodland Jumping Mice, Woodland Mice, Deer Mice, Red-backed Voles, and a Smokey Shrew in or near an owl's nest (Brinker and Dodge1993).

Northern Saw-whet Owl Pellets

Unlike pellets from Northern Pygmy-Owls, the pellets from Northern Saw-whet Owls often contain bones, fur, and/or feathers of their prey. Identifying what the birds are feeding on through pellet analysis is quite simple as long as you can identify small animals by their skulls and teeth.

Having rehabilitated several of these owls, I have accumulated a number of pellets that the captive birds coughed up. The few pellets from wild birds that I have found were roughly 24.65 x 12.54 mm. Scott Weidensaul collected pellets from wintering Northern Saw-whet Owls in Pennsylvania and found 26 pellets that ranged from 20-25 mm long and 15-20mm wide.

Typical Northern Saw-whet Owl habitat in Northern Colorado.

Adult Northern Saw-whet Owl about to begin an evening of hunting.

According to Swengel and Swengel (1992) Northern Saw-whet Owls often cast two pellets per prey item consumed, because several pellets either contained the front half or back half of a mouse or vole. Unlike Northern Pygmy-Owl pellets, pellets produced by Northern Saw-whet Owls contain a high density of bones. Unless the mouse, vole or shrew that the saw-whet has consumed is tiny, the pellets will not contain an intact skull. The few skulls I have found in the pellets were about 17.1 x 7.9 mm.

Owl pellets often contain more bones than hawk or eagle pellets. However, after the female owl has laid her eggs, the pellets she regurgitates frequently contain fewer bones due to her body's need to regenerate calcium.

When watching one of these owls cough up a pellet, it appears to be a slightly traumatic ordeal. The owl will often contort its body, head, and neck into short jerky movements. Then the owl will open its mouth, look downward, and the pellet drops to the ground. This will occasionally occur after the owl has captured its first evening meal. It seems that the pellet needs to be regurgitated before the owl can consume the fresh meal.

Chapter Six

Northern Saw-whet Owls in the Fall and Winter

There are numerous banding stations throughout the country that specialize in trapping and banding Northern Saw-whet Owls as they migrate south in the fall. Some stations trap and band these owls on their northern migration as well.

After having spent time at the Lindwood Research Station in central Wisconsin during the mid-1980s, and again in the late 1990s, banding migrating Northern Saw-whet Owls, I decided to open a banding station just south of Estes Park, Colorado. Since the fall of 2006, I've operated a banding station. I tried to model my station after Lindwood's. I used three 39.5 feet (12 m) long, 8.5 feet (2.6m) high mist nets that are either 2 3/8 inches (61mm), or 1.5 inches (36mm). These mesh nets are stretched between two poles.

With the assistance of Chris Bieker, an avid bird enthusiast, we set up a banding station along the south side of his property. The habitat is seemingly perfect. At an elevation of 8167 feet (2490 m), the area has everything the little owls seem to prefer. It contains a combination of Ponderosa Pine, Quaking Aspen, Douglas fir, and juniper, along with an active stream and small openings within the forest.

Three nets are placed in the shape of a "U" with a CD player placed in the center of the "U" broadcasting the owls' territorial call. Prior to sunset, the nets are opened, then just before dark, the player is turned on and every half-hour we walk the net line with headlamps and flashlight checking to see if any birds had been captured. When we find birds in the nets, they're extracted, placed in a small cloth bag, and transported into the house.

Once in the house, each bird is banded using a size four aluminum United States Fish and Wildlife Service leg band. Its wings and tail are measured, then the bird is placed in a small tube and weighed. To correctly weigh the bird, I weigh the empty tube first and then subtract that weight from the scale when the bird is placed within the tube and the tube is placed on the scale. Using this method, I am reading only the bird's weight.

Both wings and tail are measured by using a wing rule. This type of rule is similar to a normal ruler except this one has a small lip on it so when the bird's wrist is placed firmly against it an accurate measurement of the bird's primary wing length can be taken.

As with most North American owls, Northern Saw-whet Owls have what is called an incomplete wing molt, meaning that they do not molt all of their wing feathers each year. A bird that was hatched the same year that it is captured will have all of its wing feathers the same shade of brown.

Older birds often have two or more shades of brown wing feathers. Fresh feathers are a darker brown with worn feathers being lighter brown. Boreal Owls, Northern Saw-whet Owls, and many other species molt only a section of their wing feathers each year. The end result allows banders to see different shades of brown on the birds' wings. At times birds will be captured with two, three, and sometimes even four different brown tones on each wing, allowing us to estimate the age of the bird.

To accurately identify a bird's wing color, I extend the wing, looking at each feather. By doing this, I can determine the bird's age. Aging birds are indicated by: "HY" hatching year, "SY" second year, "TY" third year, "4Y" fourth year, and/or "AHY" after hatching year. By looking at a combination of feather color and, to an extent, shape of the outer primary flight feathers, banders can tell how old the bird is. Young of the year have pointed flight feathers, whereas adults have more rounded flight feathers (this pertains to the outer two flight feathers (the 9th and 10th only).

Moreover, there are a small percentage of birds whose age is considered AHY, meaning that they were not hatched the year they were captured, but we are unsure what year they actually hatched.

Northern Saw-whet Owl recently captured in a mist net.

Scott extracting an owl from the mist net. This usually takes a few minutes. Photo taken by Michelle Blank.

Size four leg band on a banded owl in 2006. The band fits loosely on the bird's leg and does not affect them at all.

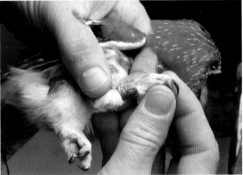

Measuring the wing of an owl. This measurement is taken using a wing ruler. The wing is measured from the bird's wrist to its longest primary flight feather.

Each bird's bill is measured as well. This may aid in determining the sex of the owl.

An owl being weighed. The weight of the can is subtracted so the weight of the owl can be read accurately.

The age of the bird can be determined by looking at the shade of the wing feathers. This is an owl that was captured the same year it was hatched. All the feathers are the same tone.

This is an owl that is considered a second year bird. It hatched the year before it was captured. Note the two tones in the wing feathers. The darker feathers are the newer ones.

Lastly, we check the amount of fat that each bird has, which is done by holding one wing open and blowing into the bird's armpit. In this way I can see how much fat the bird has. The bird's fat is white and its skin is pink. The amount of fat is estimated and recorded with a one, two, or three, with three being fattest.

The sex of the bird can, at times, be determined by a combination of the wing length and weight. Females have a longer wing length and weigh more than males. As a general rule, females weigh 93 grams or more and males weigh 78 grams or less. Males have a wing length of 120-140 mm from wrist to longest flight feather and females have a 130-150 mm wing length.

One very important thing to keep in mind when determining the sex of these and other birds is that there is a bit of overlap between males and females, meaning there is always a percentage of birds whose sex is considered unknown.

What I've found through my research is that there are a lot fewer Northern Saw-whet Owls moving through Colorado than there are in the central and eastern parts of the country.

The first three years that Chris and I trapped Northern Saw-whet Owls (2006 through 2008) we caught 11 of the small owls each year. What was interesting about that first year was 10 of the 11 birds were caught within about the same 10 foot (3.04m) section of one net and nine of 10 birds were caught in the lowest trammel (pocket) of that net, even though there were two other nets up. The following year (2007), 11 Northern Saw-whets and one Long-eared Owl were captured, with all the Northern Saw-whet Owls being caught in two of the three nets with birds being captured in all levels of the net.

The larger Long-eared Owl was captured the first night of trapping the second season. That season I tried something different. I played the territorial call of the Flammulated Owl the first few evenings to see if I could capture any Flammulated Owls that may have been migrating south. The Long-eared Owl may have come to the flam call thinking it was an adult Long-eared Owl vocalizing because both species have a similar call. Or, the larger owl may have been searching the area for an easy meal, thinking a smaller owl was actually in the area.

On a few occasions while extracting birds from the net, a second and sometimes even a third bird was calling in the immediate vicinity. As we extracted the first owl, the second was flying around vocalizing. Yet after we released the original bird, the additional owls most often stopped calling and were not captured that evening.

An owl trying to be released. After being brought outside, it takes a few minutes for the bird's eyes to adjust to the darkness before they fly off.

The same owl moved to my fingertips before flying into the darkness.

Another owl that was placed on the branch of a Ponderosa Pine so it can fly off when it wants to. Photo taken by Kim Lankford.

My thought is that, possibly, the calling birds were traveling along with the bird that was captured. Then as we extracted the first owl, the other birds watched what we were doing and moved off as we took that first bird inside to be processed. When we released the banded bird, the additional birds may have moved off together.

After each nightly banding session, the nets are taken down so not to catch any unsuspecting songbirds flying around the area during the day. Then, just before dark, the nets are replaced and opened. To attract the owls, their territorial call is broadcast beginning just after dark and continuing several hours throughout the evening. As the birds move through the area heading south, they come to the broadcasted call, eventually hit a net, and fall into one of the pockets and are captured.

Prior to release, each bird is placed outside in a cardboard box for a few minutes, enabling it to acclimate to the darkness again. Then the birds are released and they instantly fly into the night.

A Long-eared Owl that was captured during the 2006 banding season. Photo taken by Kim Lankford.

Banding Northern Saw-whet Owls

Some banding stations capture several hundred birds each fall. According to the Ned Smith Center for Nature and Art in Pennsylvania (www.nedsmithcenter.org), the numbers of Northern Saw-whet Owls captured each year varies from one year to the next. In 2007, banders netted 904 Northern Saw-whet Owls, their best season to that point, with 27 birds captured that were originally banded elsewhere.

However, in 1999 that center and others like it captured large numbers of these little gems. In fact, the Ned Smith Center captured more than 850 Northern Saw-whet Owls that year. Throughout the state of Pennsylvania, banders trapped 1,250 saw-whets in 1999.

The Beaverhill Bird Observatory in Edmonton, Alberta, captured 145 Northern Saw-whet Owls in 2002 and 151 in 2003. The nets were set up for between four and six hours each evening depending on the month and the weather. In 2007, the Prince Edward Point Bird Observatory (PEPTBO) banded 1,518 saw-whets that fall season. White Fish Point Bird Observatory (WPBO) banded 269 Northern Saw-whet Owls in 2007 and 314 in 2006.

On the other hand, banders in Edewold, Saskatchewan, banded 207 owls in 2007, yet only 140 in 2006. All of this shows that Northern Saw-whet Owls move through different areas of the country in different numbers each year.

The Thunder Cape Bird Observatory (www.tbfn.net) in Ontario, Canada, has been banding these owls for several years as well. Birds are trapped from mid-September through late October. From 1994 through 1998, banders at the observatory banded 1,273 Northern Saw-whet Owls and have had several birds recovered.

Some of their banded birds ended up 260.4 miles (420 km) away in Stevens Point Wisconsin, yet others were recovered at Hawk Ridge in Minnesota 168.44 miles (272 km) away. One of the owls was originally banded at the observatory in Canada on 28 September 1991 and recaptured in Stevens Point, Wisconsin, 3 October that same year. Another similar scenario was of a bird banded 13 October 1994 and recovered 6 November 1994 in Cedar Grove, Wisconsin, 334.8 miles (540 km) away.

On 2 October 1995, the observatory banded 53 birds and had three recovered several miles from the site. One was recovered at the Hawk Ridge banding site in Minnesota, another in Stevens Point, Wisconsin, and the third in Whitefish Point, Minnesota. The three birds traveled an average of 16+ (27 km), 17+ (28 km), and 18 + (30 km) miles each evening to reach their recapture destination.

In 2007, banders in Indiana captured a banded owl that was originally banded in Wisconsin 22 nights earlier. That particular individual had moved about 400 miles (660 km) in 22 nights, an average of 18 miles per night.

The majority of the birds captured each year are either adult females or juveniles. Banders in the fall routinely capture fewer males than females, which tells me that the males are staying on or near their territories throughout the year, while the females and young head south.

As a result of an increased number of banders trapping Northern Saw-whet Owls during fall migration, I did a bit of research to see how many birds had been trapped and subsequently banded. Between 1955 and 1969, banders throughout North America banded 4,802 Northern Saw-whet Owls. Then I checked the Bird Banding Laboratory's website (www.pwrc.usgs.gov/bbl/) and found that from the inception of the banding laboratory in 1914 through 2002, there were 94,082 Northern Saw-whet Owls banded and 1,785 encountered. Then I found that from 1914 through 2004 banders banded 113,726 of these little owls and had 2,092 encountered. Finally, between 2002 and 2004 banders banded 19,644 Northern Saw-whet Owls and encountered 307.

An encounter is any time the bird's band number is read. In some cases, the birds may have been found dead and in other cases they may have been recaptured and released unharmed.

Northern Saw-whet Owl perched within a Ponderosa Pine.

I found this owl in early March 2007, in what ended up being its nest that year.

I found this owl perched in a campground on the eastern plains on Colorado.

This owl was vocalizing well before dark one February afternoon.

The Northern Saw-whet Owl   141

Not All Northern Saw-whet Owls
Are Migratory

Within their winter range, Northern Saw-whet Owls can occur in large numbers locally. There are numerous records of these birds in northern latitudes during the winter months. Bent (1938) wrote that these birds are found as far north as New England and Oregon in winter. During the winter of 1878, more than 21 of these owls were taken near Princeton, New Jersey, by a Mr. Scott. Similar situations occurred that same year in Oregon as well (Bent 1938).

During the winters of 1955-56 and 1956-57, Mumford and Zuni (1958) discovered small groups of Northern Saw-whet Owls wintering in or near the Edwin S. George Reserve in Livingston County, Michigan.

The owls were located by systematically walking the reserve. They were most often found perching on a horizontal branch within a foot of the trunk, with some individuals being located after researchers had located white wash, feathers, and/or pellets under branches.

Several birds were banded and color marked so researchers could discern individuals. By marking these birds, some interesting information was found. For instance, the researchers found two different individuals roosting in the same tree on different days. They also found that individuals would perch on the same perch on and off throughout the winter as well as finding an individual that would roost in a single area for several months before moving on.

In 2007, I came across two of these little owls in Estes Park, Colorado. Both birds were found on private property: one that had hit a window in late January (discussed in Chapter Five) was an adult female and the other (sex unknown) was an adult bird as well and was perched near a bird feeder in early February.

That owl was at Scott and Julie Roederer's house, perched in a spruce tree. Scott originally noticed the owl that morning, as the songbirds that normally congregate at his feeders, i.e. juncos, nuthatches, chickadees, etc., were voicing their displeasure as they noticed the predator. After a few minutes the birds went about feeding and continued doing so throughout the rest of that day as if the owl weren't even there. The owl made no attempt to attack any of the songbirds that day, and the birds seemingly forgot about it. The owl was apparently hunting near the house the previous evening and decided to roost in what it thought was a quiet place. It was obviously unaware that the area that it picked for roosting was going to fill with songbirds during the day. As nightfall approached, the owl moved off and was not seen again.

With the two above-mentioned sightings as well as other individuals that have been heard vocalizing in late January (in Drake, Colorado), it appears to me that at least some of these individuals may be nonmigratory, assuming that the vocalizing birds were territorial males. If the birds do in fact migrate, they can't be moving very far if they begin calling so early each year. They seem to call on and off during January and February however, by March and April, the males will often be calling with more serious intentions.

Every time that I've found one of these owls on its daytime roost, the bird was either perched quietly in a relaxed position with its eyes closed perched next to the trunk of the tree. Or it decided to conceal itself by making a couple of simultaneous changes. When concealing itself, the bird would straighten up, raising the outer corners of its facial disk, making it appear to have ear tufts. It would also raise the rictal bristles around its beak, tighten its body feathers making it appear long and thin, and pull one wing over its body. Often this action will make the bird undetectable to birders or potential predators.

The bird would take this action as I walked near it; as I moved away, the owl would loosen up, transforming itself back into its original relaxed position. If I would return to the bird, or if it was an individual that many people had already looked at, the owl would often remain in its relaxed position as it was watched. Apparently, when the bird feels no threat, it will remain calm, making no attempt to conceal itself.

This beautiful bird was perched within a few feet of several active
bird feeders as the songbirds flew to and from the feeders all day.
There was no interaction between the predator and the songbirds.
© Gary Mathews

The Northern Saw-whet Owl   143

Chapter Seven

Rehabilitation of
Northern Saw-whet Owls

In order to become a wildlife rehabilitator in Colorado, you first have to obtain the proper forms from your state and federal Division of Wildlife office, fill out the paperwork describing what birds and/or animals you are planning to rehabilitate, and then find a rehabilitator that agrees to sponsor or assist you if you need help or have questions. Then you need to find a veterinarian who will agree to assist you if needed. Also you must construct a flight cage for the birds to exercise in, after which the local Division of Wildlife officer needs to inspect it.

All of this is to be completed before you can legally accept your first injured creature. The majority of wildlife rehabilitators do this work with no financial assistance from anyone. All the rehabilitation costs come out of the pockets of the rehabilitators themselves.

I acquired my first injured Northern Saw-whet Owl in 1994, after it had been attacked by a cat and taken to a local veterinarian. The doctor called and explained that he had an injured Great Horned Owl, which he was going to keep over night for observation and would I be able to pick it up the following morning.

I arrived at the vet's office the following morning and, upon entering his waiting room, I explained to his nurse who I was and what I was there for. A few moments later, the doctor entered the waiting room. After exchanging formalities, he explained that he had found no apparent injuries; however, the bird was unable to fly.

He went into his back room and came out wearing leather welding gloves, carrying a small box about 10 inches (5.40 cm) cubed. I said, "Is the owl in that box?" He said, "yep." I told him that if the bird fits in that box, it's definitely not a Great Horned Owl.

He placed the box on the table and I slowly opened it up to find the bright yellow eyes of an eight-inch (23.32 cm) Northern Saw-whet Owl looking up at me.

He explained that the bird had no visible injuries, but was unable to fly. I asked if he knew how the little owl became injured. He explained that the bird was apparently perched in a small juniper in someone's yard and their cat found it and apparently thought it might be something fun to play with it. However, the cat quickly realized that the owl wasn't as much fun as originally thought, and the cat released the owl, but not before permanently injuring the owl. The owl spent the next seven years at the Birds of Prey Foundation before it died of West Nile virus.

Other Injured
Northern Saw-whet Owls

Since receiving that first owl, several of these little gems have found themselves in my care. Some had injuries, yet others just ended up in a precarious predicament and needed some temporary assistance.

Virtually every year that I've rehabilitated birds, at least one and sometimes more of these little owls have come my way. Most are found during migration, yet others during the nesting season, and still others are brought to me in the winter.

One little bird was run over by a truck, well sort of. A friend of mine was on his way to work one evening, when he saw, on the yellow line of the road, a small bird. His car straddled the bird, so he stopped, backed up, and saw a little owl perched on the centerline of the road. He picked the bird up, placed it in a box, and brought it to me.

I checked the owl for any injuries, fed him a mouse, and left him alone. I found no injury on the bird, so

I received this little owl after it had broken its wing. Unfortunately, the wing did not heal correctly; therefore, the bird remains at the Birds of Prey Foundation in Broomfield, Colorado.

From top:
A pair of owls at the Birds of Prey Foundation.

Three young owls in a pet carrier at the Birds of Prey Foundation.

An owl inside my flight cage. Note the vole that the owl has killed inside its flight cage. I don't think the vole would have entered the cage if it had known an owl was inside.

The Northern Saw-whet Owl   145

the following evening I placed a band on the bird's leg, took him back to the vicinity in which he was found, and released him. The owl flew into the woods and has not been heard from since.

I believe the owl made an attempt to catch a mouse or other small animal as it ran across the road. Upon missing the creature, the owl stopped momentarily on the road just as the car passed. Being temporarily blinded by the headlights, the bird remained motionless, allowing the person to pick it up.

I acquired another owl early one morning after a couple found one standing on their walkway unable to fly. I met the couple at their home and, after we took a short walk, the owl was found under a small spruce tree. I placed a small towel over the bird, picked it up, placing it into a pet carrier, and went home.

At home I put the pet carrier in my office and removed the owl from it. The bird was unable to open his wing because there was dried blood on its wing and breast. Evidently, the owl had gotten blood on its wing or chest from something it had consumed. While the bird was roosting during that day, the blood dried and the bird was unable to open its wing and fly.

The owl is perched on the top of the picture frame, with the stuffed quail in the lower left corned.

Once in the office, I washed the blood off the bird and tried to place it back into the carrier. However, as I attempted this, the owl exited the carrier and flew around the room, landing next to a stuffed California Quail that I have on a shelf. I got my camera and took a few photos before capturing the owl. By this time it was about 9:00 p.m., so I took the owl down the street to an area that is suitable saw-whet habitat and released it.

Still another Northern Saw-whet Owl was brought to me one spring morning after it was also found unable to fly. This bird was truly injured. It had a broken humerus, which is the bone that stretches from the elbow to the shoulder.

I placed the two ends of the bone together and wrapped the wing, then wrapped the wing to the bird's body. The owl was placed in a pet carrier with a soft towel on the bottom along with a dead mouse and a shallow dish of water.

For the next 10 days, I kept the bird quiet and well supplied with mice. On the 11th day, its wrap was removed and after a few days, the owl was placed in a flight cage to exercise. Later that fall, the owl was released into an area where Northern Saw-whet Owls had previously been heard.

Two different owls were brought to me after they hit windows. Just as diurnal birds hit windows during the day, nocturnal birds occasionally hit windows after dark. The owl was apparently hunting near someone's house and hit their picture window. The homeowners heard a thud on the window and walked out onto their deck to find a little Northern Saw-whet Owl lying unconscious.

A few minutes later I arrived at their house and found that they had already placed the owl in a small box. I could hear the owl thrashing around within the confines of the box. It was not very happy being confined. I took the bird home, kept it for three days, and after finding no injuries, I released it where it was found.

Another window crash occurred several years later, but this owl hit a window at 10:30 a.m.! These birds are supposed to be nocturnal, or so I thought!

My pager went off one morning and after calling the number on the screen, a voice on the other end explained that she had a baby owl that had just hit her window. I explained to her that it was probably an adult Northern Pygmy-Owl because there are no baby owls in Estes Park, Colorado, during early January.

After about 10 minutes, the girl arrived at my house. Upon opening the box, I saw that it was, in fact, a Northern Saw-whet Owl! I was quite surprised because it was clear and sunny that morning, which is another reason why I thought the owl, would be a pygmy-owl, because Northern Saw-whet Owls are supposed to be inactive during the day.

Apparently, the owl was roosting in a spruce tree in the girl's yard and its hiding place was discovered by magpies. The corvids must have scared the owl from its day roost and it flew into the window trying to escape the onslaught.

I placed the owl in a pet carrier indoors, along with a water dish, a few dead mice, and a box perch. The owl remained indoors for 10 days, after which time it was placed in my outdoor flight cage.

Before the owl went into the cage, I threw some bird seed on the floor of the cage to entice mice to come into the cage. This worked perfectly, because after the first night, the owl was no longer eating the mice I put in the cage, but was catching wild mice on his own. The owl was released after it was in the flight cage for a week.

Flight Cage Description

Since becoming a bird rehabilitator, I have built three different flight cages for the injured birds in my care, with each cage having been designed larger than the previous one. When creating a cage to house injured birds, one needs to determine what species will be housed within the cage and how the cage can best aid in the rehabilitation process.

For example, the first injured bird that I received was the above mentioned Northern Saw-whet Owl (the one that was attacked by a cat). When a bird such as that one is unable to fly, the cage must have several branches throughout, allowing the bird to move freely through his apartment, getting exercise. Make sure that as least one of those branches reach the ground, that way, in case the owl ends up on the ground for any reason, it can easily climb to a more comfortable perch. Birds such as Northern Saw-whet Owls feel safer off the ground than on the ground.

The flight cage should also have ample ventilation which is most often created using vertical slats along the sides and horizontal slats across the top. The vertical slats are placed, one half inch apart so the owl and

A Northern Saw-whet Owl perched near the corner of my flight cage a few days before its release.

subsequently other birds can get sunlight and fresh air, but will be unable to escape. If your cage is to house larger birds, the slats can be up to an inch or more apart depending on what size birds are in the cage.

Furthermore, I make sure the cage has at least two sides that are enclosed and the rest slatted. When small owls are being housed in my facility, I place a small water dish along with an ample supply of mice for them to feed on. The facility also has a nest box inside, just in case the owl wants to conceal itself during the day.

When I received my first Northern Saw-whet Owl, I was extremely inexperienced as far as rehabilitation goes. When I brought that bird home, my original cage was eight feet square (2.43 m) and only had two perches in it. The cage was completely enclosed with the exception of a small two-foot (60.96 cm) square slated window. It had two perches in it, one that extended from the ground to a nest box and another that stretched from one side of the cage to the other.

After that Northern Saw-whet Owl was in my possession for a few days, I noticed its wing began to open at the wrist. So I contacted The Birds of Prey Foundation in Broomfield, Colorado, and talked to its founder and president, Sigrid Ueblacker, who invited me to her facility to look at her cages.

The following morning, Sigrid showed me her Northern Saw-whet Owl cage. I was astonished with the amount of thought she put into that cage and the other cages on the property. Her Northern Saw-whet Owl cage is roughly 30 feet (9.15 m) long, 20 feet (6.10 m) wide and 15 feet (4.57 m) high. It has a wall with a door in the middle of the cage that can be closed to create two smaller cages if needed.

Within the cage are several nest boxes. Some have perches leading to them; others do not. The walls are covered with small branches, which allow the flightless owls to move throughout the cage. There are even live spruce trees inside both the front and back sections of the cage allowing the birds to hide in the trees if they want. Parts of the cage have a covered roof, yet other parts are open to the elements. There are two large water dishes and several stumps on the ground as well.

The combination of all these things gives the birds choices. They can sit in the sun during the day if they want or they can hide in a nest box, or even take a bath if they'd like. If two individuals don't get along, the cage is large enough for those individuals to stay away from each other. Conversely, if two birds like each other, they can be alone within the cage as well. I learned more that afternoon than I would have ever expected and it changed my concept of rehabilitation.

On the way home that afternoon, I began thinking about how to incorporate some of the branches to enhance my flight cage. The following day, I went into my cage and nailed several small perches to the insides of the cage, making sure that the branches were small enough that the owl could wrap its toes entirely around the smallest ones. Birds prefer to perch on branches that enable them to wrap their toes completely around the limb.

What I immediately found was that the owl walked to the highest point in the cage, which was also the darkest one. There were several instances when the owl would perch at the window at night, which made me begin to think about constructing an addition onto the cage that would be entirely made of wood slats.

A few weeks later, a few friends of mine and I built the addition onto the flight cage. The walls were constructed completely of vertical slats. The roof was slatted as well, which gave the owl, and all subsequent birds, the ability to be out in elements if they preferred.

During the day, the owl would either be inside the nest box, or perched on the tip of the highest branch in the cage. I knew that the bird was much happier now and I felt the larger room and extra perches made the bird happier.

As I approached the cage to clean it one morning, I saw the owl drinking from the pool I placed in the cage, which was the first time I had seen a Northern Saw-whet Owl drink water. Apparently, these owls seldom drink water because they are able to get the majority of their water intake from the mice they consume.

After seven months, I received an injured Prairie Falcon and, because I only had the one cage, I contacted Sigrid and she agreed to take the Northern Saw-whet Owl. The bird was in her facility for six years before it died of the West Nile virus.

I now (2008) have a flight cage that is 16 by eight feet and eight feet high (4.87 by 2.43 m), 2 feet (.60m) of which is completely enclosed, except for an opening that enables the birds to get into the open area, which is completely slatted, tops and sides. There are several perches that I periodically change around and add a different sized water dish depending on which species is inside. I have also placed an electric fence around the outside of the facility, which keeps the bears and raccoons out.

Mortality and Longevity

Before Northern Saw-whet Owls were federally protected, both adult birds and their eggs were routinely collected for various personal collections and/or museums. Fortunately, this practice was stopped after the birds were federally protected.

A number of these little owls find themselves in deadly situations each year. For instance, nestlings have perished within their nests due to starvation, insect infestation, and predation by mammalian intrusion.

Once out of the nest, the inexperienced young owls are occasionally captured by hawks during the day and owls at night. Screech, Barred, and Great Horned Owls have been known to dine on the smaller Northern Saw-whet Owl. Occasionally Northern Saw-whet Owls end up caught in houses still under construction and perish from starvation. Some even crash into windows at night in the same manner as diurnal birds hit windows during the day.

The Birds of Prey Foundation in Broomfield, Colorado, began receiving a few injured Northern Saw-whet Owls in the early 1980s, with a greater number of admissions by the late '80s. This increase in admissions is directly related to the increase in house and shopping mall construction. A number of the injured birds were found on construction sites inside houses and buildings that were under construction.

That same facility began losing owls to the West Nile virus in the summer of 2003, yet by 2005 no Northern Saw-whet Owls were lost to the disease. Some birds may actually become immune to the virus, which could explain the lack of cases in 2005.

A fair number of these little owls are found dead along roadsides each year as well. In the spring of 2005, while traveling just outside Loveland, Colorado, I found a saw-whet owl dead on the side of the road, apparently hit by a car. I found another dead owl hit by a car on HWY 287 in Boulder County, Colorado, in 2007.

Another owl banded in Bigfork, Minnesota, on 26 October 2006 was killed striking a motor vehicle in Bismarck, North Dakota, on 14 November 2006. Unfortunately, this occurs with much more frequency than any of us would like.

By far the most upsetting cause of death that I've come across was on 20 July 1999. While driving to work that morning, I noticed a fairly large bird hanging lifelessly from a single tine of a barbed wire fence. Next to the fence was a wax current bush. My guess is that the owl was perched on that bush and, after seeing a potential meal, it took flight, got stuck in the fence, and after a struggle expired.

On the other hand, according to the Bird Banding Laboratory, the oldest wild Northern Saw-whet Owl on record is 10 years, four months old as of 2005. Kay McKeever of the Owl Foundation told me of a captive Northern Saw-whet Owl that she had that lived for 16 years and, interestingly enough, in its later years, the bird's plumage whitened as the bird aged.

This poor bird was killed after it got stuck in a barbed wire fence.

These little owls on occasion get hit by cars as they migrate in the spring and fall.

Portraits of both Northern Saw-whet and Boreal Owls.

Part Four:

The Boreal Owl

Aegolius funereus

The Boreal Owl

Chapter One

My First Encounter

The Boreal Owl is the largest of the cavity nesting mountain owls. Throughout their western range, they are found within the higher spruce-fir forests above 10,000 feet (3050 m) in Colorado, as low as 5182 feet (1580 m), and in Idaho (Hayward et al. 1987). Boreal Owls are located from February through May as the males solicit females. Their snipe-like winnowing call, at times, can be heard during the day but is most often heard after dark. Boreal Owls can also be located in the fall, specifically September and October close to or during full moon nights (Palmer and Rawinski 1988) when they often respond to a playback of the territorial call.

For me, the Boreal Owl is the most difficult mountain owl to study because within the areas that I research, there appears to be a very small number of birds. Boreal Owls are also found at such high elevations within RMNP that during their spring courtship the areas that the birds are in often has an excess of four feet (1.21m) of snow on the ground, making movement through the woods quite difficult.

My first Boreal Owl encounter was in1994, when I was assisting Dr. Ron Ryder, emeritus professor from Colorado State University, with his research in the Cameron Pass region of northern Colorado. Ron and Dave Palmer, a wildlife biologist, along with other researchers and volunteers had been studying the owls in that area since 1979.

Through their research, they were able to locate nesting birds in both natural cavities and nest boxes (that they had put in place). They had placed radio transmitters on some individuals to help discover territory sizes and discover what the birds fed upon. Road censuses were also conducted in the spring to identify numbers of birds in that area along with locating potential territories that they could later search for nest sites.

On 10 March 1996, I assisted Ron with one of these road censuses. It was about 6:00 p.m. when we arrived at a small dot on the map called Gould, Colorado. The objective was to drive east to Chambers Lake over Cameron Pass (elevation 10,276 ft. (4318 m)), stopping every half-mile to listen for any calling birds. If none were heard, we would play a recorded Boreal Owl call and document what would transpire. That particular evening we heard nine different birds.

One of the normal stops along the route was the Moose Visitor Center, just between Chambers Lake and Gould. We pulled into the parking lot, turned the car off, and stepped onto the pavement to hear an owl calling in a spruce tree near the road just a few yards from us. Ron suggested that we pick two different spots that would allow us to triangulate on the calling bird. As we did this, Ron said "shine your light to where you think the bird is and I'll shine mine to where I think the bird is."

We did this and "*voila,*" there was a Boreal Owl right in front of us. We watched it vocalize for a few minutes, during which time I set up my camera – complete with tripod and flash – and began taking pictures. I was ecstatic because, as a bird watcher, this was a "life bird" for me. I was so impressed, first with how many birds we heard that night, and then actually seeing one made it an extraordinary evening.

Boreal Owl bobbing its head. Boreal Owls often bob their heads when they are trying to hear or see better. © Bill Schmoker

Earlier that evening, when Ron called me to ask if I would assist him with this project, my first question was, "What are the chances of actually seeing an owl?" He said, "You're definitely going to see one, so make sure you bring your camera."

As we made our way west on Highway 14 toward the pass, Ron said, "I have to confess something to you: the odds of actually seeing a Boreal Owl are very slim, but we're virtually guaranteed to hear some. I just needed someone to help me." I told him, "I would've come along regardless of whether we would see birds or not."

I was just as interested in helping him as I was in seeing the bird. I have always said that free education is priceless. As we made our way to the pass, I told Ron that the area around the pass looks very much like the region within the Hidden Valley area of RMNP. He said that he had always thought that area of the park should be good for Boreal Owls.

As we were driving to the pass, we talked about why he began studying the owls in Colorado. He explained to me that prior to 1965 there were only four Boreal Owls documented in Colorado, with the first being obtained in 1896. Then in August 1963, Baldwin and Koplin collected a juvenile female on Deadman Mountain in Larimer County. He also said that throughout 16 years of roadside counts along Cameron Pass he found between zero and 27 calling owls. He suggested that this may be due to the volume of food available to the owls during those years, i.e. the more food available, the more owls.

A few days after my Cameron Pass experience, I talked a friend of mine into going to the Hidden Valley area of RMNP to search the area for Boreal Owls. That particular section of the park was at one time a downhill ski area. By the time I began searching the area for Boreal Owls, the only remnants of a ski area were the unused ski runs covered in snow.

On 31 March 1996, we arrived at Hidden Valley at 8:15 p.m. and began walking up one of the unused runs. A few yards up the trail, a Boreal Owl began calling. I mimicked the owl's territorial call and a few moments later the bird was within a few feet of us. It was quite an experience for both of us to see a wild owl just a few feet from us.

Boreal Owl habitat within Rocky Mountain National Park

What's In a Name?

The Boreal Owl, also called the Tengmalm's Owl in Eurasia and historically called the Richardson's Owl in the earlier days in North America, got its name (at least in North America) from the fact that it is found throughout the boreal forests of the world. Its scientific name *Aegolius,* from the Greek *aigolios,* a kind of owl, and from the Latin funereal referring to a bird of bad omen because of its call sounding like the "wailing of the dead" (Cous 1882).

Tengmalm's Owl is named after Pehr Gustaf Tengmalm (1754-1803), a Swedish physician who published papers on both medicine and ornithology. In 1788, J.F. Gmelin named the owl (*Strix tengmalmi*) because he believed that Tengmalm was the first to distinguish it. Unfortunately, a Professor Rudbeck had produced color plates of the owl in 1746, long before Tengmalm was born, but due to the scientific name being used for so many years, the Tengmalm name stuck with the owl.

Distribution and Range

The Boreal Owl, as its name implies, inhabits the boreal forests of North America, Europe, and Asia. Within North America, they are found from the tree line of northern and central Alaska, through the central Yukon south through central and southern British Columbia, and east through northern and central Saskatchewan, Manitoba, northern and central Ontario, central and southern Quebec, and New Brunswick in Canada. Within the continental United States, the owls have been found breeding, or least present, in the higher mountains of Washington State, Oregon, Idaho, Montana, Wyoming, Colorado, and New Mexico (Johnsgard 2002, U.S. Forest Service 1994, Stahlecker and Rawinski 1990). The species has nested in Minnesota as well (Matthiae 1982).

In winter, the birds often wander south of their breeding range and end up in areas such as Wisconsin, Michigan, New York State, New England, Pennsylvania, and North Dakota (Johnsgard 2002).

The green area shows the breeding range of Boreal Owls in North America.

Chapter Two

Anatomy of the Boreal Owl

The Boreal Owl, genetically speaking, is more closely related to the Northern Saw-whet Owl than it is to any other North American owl. The two species are in the same genus, yet different species. The Boreal Owl is however larger and darker than its smaller cousin and is often found at higher elevations and farther north. Boreal Owls are about 10 inches (25.5 cm.) from head to tail and have a wingspan of 21 inches (53.5 cm) with females being quite a bit larger than males (approximately 1.05 times larger Johnsgard 2002).

Boreal Owl and Northern Saw-whet Owl perched side by side at the Birds of Prey Foundation.

Like the Northern Saw-whet Owl, the Boreal Owl has long, rounded wings and a comparatively long tail with white oval spots. As the owl perches, its tail extends past its folded wings, which is unlike the Northern Saw-whet Owl, whose tail and folded wings are about the same length.

The Boreal Owl has very long and broad wings, with the leading edge of the ninth and tenth primaries fringed like a comb, allowing air to flow through the edges of its feathers, muffling the sound of the bird in flight, enabling it to fly through the forest virtually silently. The ninth and tenth primaries are the outer two flight feathers of each wing.

The Boreal Owl has a relatively large head with a pale bill and small yellow eyes. A border of dark feathers surrounds its white facial disk. Its forehead, top, and hind head are dark brown covered with white spotting. Its hind head has two large white spots on the nape that somewhat resemble eyespots, but they

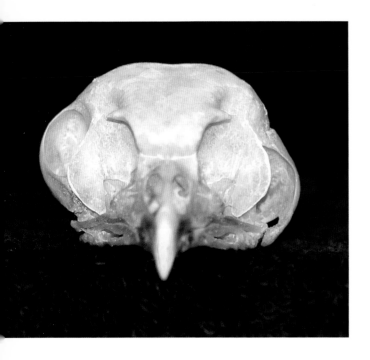

are not as pronounced as they are in the pygmy-owls. The bird's wings and back are a dark brown with light spotting as well. Its breast and belly are white with gray-brown, and in some individuals, rusty-brown vertical streaking.

The owl has feet feathered to its razor sharp talons and in some individuals the feet are so heavily feathered that they resemble a mammal's paw, with only the pads of the feet and talons being without feathers.

The ears of this owl are extremely asymmetrical, with the right ear on average 6.3 mm higher than the left (Norberg 1968), enabling sound to reach the ears at two slightly different intervals, allowing the bird to pinpoint and capture any mouse, vole or bird that emits a sound. Due to the degree of asymmetry of its ears (most asymmetrical of any North American owl), the Boreal Owl may have the best hearing of any of the North American owls as well.

The Skull of a Boreal Owl. Note the asymmetry of the ear openings, which are on the sides of the skull.

Vocalization

The territorial call of the Boreal Owl (males only) is a series of winnowing notes described as the "staccato song" by most researchers. It sounds like a series of *boo.boo.boo.boo.boo.boo.boo* that rises slightly at the end. This call resembles the aerial display sound emitted by the Wilson's Snipe. However, the snipes' call is a structural sound created by the stiff outer tail feathers shivering in the air as the bird dives during its courtship display, where the Boreal Owl's call is produced vocally and often carries 1.5 km away and sometimes even farther (Hayward 1993).

This primary call is often given within 30 feet (100m) (Hayward & Hayward) and at times within three feet (10m) (Meehan 1980) of a potential nest site. Calling birds vocalize from February through June depending on where the birds are within the range of the species. As with other species, Boreal Owls tend to reduce vocalizing after a pair bond is established. Hayward's (1993) observations are that the bird's primary call continues until the female occupies the nest prior to egg laying. In addition, by April the males that are still vocalizing are, for the most part, unpaired males who by this point generally begin a reduction in calling frequency.

I've found that some males begin calling just after sunset, yet others don't begin calling until close to 10:00 p.m. each evening. I presume this may be determined by the number of females in the area. Fewer females may reduce calling each season. On several occasions the male(s) in RMNP would not begin calling until 10:00 p.m. or later. I was unable to locate a nesting bird within the park, which may have meant that the birds were unable to locate females.

The variation in the time of the bird's vocalizing each evening may have a correlation to the availability of prey. For example, maybe the male begins vocalizing only after he has procured a morsel to give any female that comes its way. If his intention is to show the prospective mate how adept he is at catching dinner, he better have something to offer her. So maybe the males don't begin vocalizing until after they have something to present to a female.

On 22nd March 2005, Gary Mathews, a birding friend, and I were in RMNP searching for Boreal Owls. We arrived at the Hidden Valley area of the park just before 9:00 p.m. We began walking up one of the old ski runs, periodically playing the territorial call of the owl as we hiked. There are several old runs in that area, so we picked one where I had previously heard owls.

Hearing no owls and having no response to our soliciting, we moved up the slope, stopping periodically to play the call for a few minutes. At 9:17, an owl called on the adjacent slope. We moved to the area where the bird was heard, but found no bird. After waiting several minutes, I played the recording very softly for approximately 30 seconds, trying to entice the bird to respond, allowing us to identify the bird and gain some insight into whether or not the bird had chosen a nest site close by.

After several minutes, the bird made no sound, so we decided to go home. As we began to move off, the owl began calling just behind us. We turned around and shone our flashlights toward the calling bird. The owl was roughly 20 feet (6.09 m) above us, perched on a dead spruce branch.

The bird didn't seem to care in the least that we were standing just a few feet away. After a few moments, it began giving its distinctive winnowing, and it uttered a call that I had not heard before. It made a two and sometimes a three-note call with the first two notes sounding somewhat like the call notes of a Cassin's Finch, with the third note (not always being very audible) being a soft click.

During this calling period, he would constantly look into the woods, presumably toward its nest cavity, but I have no way of knowing that. On several subsequent occasions that spring, I heard the bird calling from that same general vicinity; however, I was unable to locate its nest.

As with many other owls' vocalization, the Boreal Owl's call is often ventriloquial, appearing to be coming from one place, when in fact the owl is in an entirely different spot. At times, I have been directly under a calling male watching him vocalize, and yet the direction he was facing determined where the call seemed to be coming from.

While vocalizing, it often perches on an exposed limb of a live tree at times leaning forward when calling. Occasionally it would even raise the upper corners of its facial disk, making the top of the head form a kind of "M".

As he was calling in front of us, I began making soft squeaking sounds, to see how the bird would react. He began turning his head upside down and vigorously pumping his head side to side, then up and down, presumably trying to locate what it may have believed was a potential meal.

Captive Boreal Owl.

Other Boreal Owl Calls

When the staccato song is extended, it is termed the *Prolonged Song* (Hayward 1993) and is uttered by the male as courtship begins and continues through the beginning of incubation. This call consists of a long trill given more softly than the bird's primary call, but lasting up to a minute. It is given near the nest as the female enters the area. According to Meehan (1980), the male at times makes repeated flights between the nest and the female, singing this particular song.

Other calls emitted by the male include a *Skiew* call, which according to Bondrup-Nielsen (1978) may be an anti-predator call given when someone, or an animal, is near the nest.

A *Moo-o* or a *Hooh-up* call may be a food delivery call described as a kind of a creaking of a tree in the wind or even a baby crying.

The *Chuck* call is a harsh, commonly modulated call that the female gives while in the male's territory in response to his primary call. The *Chuck* and *Skiew* calls are the most frequently heard calls given in the fall as a response to a playback of the territorial call (Rawinski per.com).

The *Peeping* call is a soft vocalization given by the female throughout the breeding season, described as a "pure tone starting with a rapid increase in pitch" (Bondrup-Nielsen 1978) followed by a continual section ending with an accelerated drop in pitch.

When Do Boreal Owls Begin Vocalizing?

Research suggests that the time to begin listening for Boreal Owls, at least in Colorado, is from January through June (Ryder et al. 1987 and Palmer et al. 1988). I've found them vocalizing quite consistently in late March and April, at least in Rocky Mountain National Park, giving their staccato call. Palmer (1987), in Colorado, wrote that singing males peak in late April. Birds calling in June may be unmated males, or first time breeding birds (Hayward 1993).

In Colorado, courtship calling can last over 100 days with an average being 26 days (Palmer 1986).

Winter Boreal Owl habitat in Northern Colorado.

Chapter Three

Courtship

As with other nocturnal owls, the courtship of Boreal Owls is a bit hard to identify; however, there are some accounts of their courting activities. According to Seton (1911), in Bent (1938), the male flies circles around his potential mate, giving a soft but high-pitched bell-like *ting,ting,ting,ting,* which rises and falls at about two *tings* per second. Seton listened for 20 minutes before going to sleep, but continued to hear the call a few times throughout the night.

The owl's courtship begins with the male's staccato call ringing throughout the forest. This call contains roughly 10 to 16 toots per bout. Then, when a female approaches, his call can continue for over two minutes without a break. At times, when the male hears a female calling in response to his solicitation, he will fly directly into the nest cavity and begin calling from it (Bill Lane, a Boreal Owl researcher, personal comm.).

At other times, when the female is present, the male will call, giving his prolonged staccato call and flying from the female to the nest site, then back and forth, attempting to entice her into his penthouse. Once they share the same address, the courtship feeding begins, which can continue for up to three months prior to the commencement of nesting.

Nesting

Throughout the western United States, Boreal Owls' nests are most often found at least at 9500 ft. (2900 m) in mixed mature spruce-fir forests, including Subalpine fir and Engelmann Spruce. In Minnesota, the owls were found in mature Black Spruce, Balsam Poplar, and White Cedar (Matthiae 1982). In Alaska (Meehan and Ritchie 1981), the owls were found in mixed stands of White Spruce, Paper Birch, and Quaking Aspen.

Adult Boreal Owl looking out of its nest cavity.

Boreal Owls have been found nesting in abandoned woodpecker cavities excavated by both Northern Flickers and Pileated Woodpeckers. They've also been found nesting in nest boxes made to the proper dimensions and placed at the proper height on the tree. The dimensions for a nest box that I use are eight inches (20.4 cm) wide and 16 inches (40.7 cm) deep with a three-and-a-half inch (9 cm) entrance hole. I like to use at least one-half inch thick wood and place an inch or so of straw on the bottom of the nest box so that the eggs won't roll around while the female is incubating, because the female owl won't add any nest material to her nest cavity.

In Bent (1938), R.T. Tufts (1925) discovered a nesting owl in an abandoned flicker cavity about 10 feet (3.04m) from the ground, which apparently was about the height of that tree. The bird's face filled the entire opening of the cavity and as the bird peered from its entrance, and as the bird exited, it apparently had to "hitch from side to side before exiting."

Also a Dr. Fisher (1893b) (Bent 1938) claimed that "it [the Boreal Owl] is very partial to the old holes of the Pileated Woodpecker, which seem to be just the right size and shape to suit its fancy."

Palmer and Ryder (1984) found four nests from 1981 through 1984, two of which were in Lodgepole Pine with the other two in Engelmann Spruce. Matthiae (1982) found a Boreal Owl nesting in a nest box that was originally erected for waterfowl. It was just a bit higher than 12 feet (3.65m) from the ground. Hayward (et al. 1993) found the owls nesting in Quaking Aspen, Engleman Spruce, Douglas fir, and Ponderosa Pine as well. Holt and Ermatinger (1989) found the first nesting Boreal Owl in Montana in a Subalpine fir.

Apparently, this species seldom uses the same nest for more than a year at a time. However, different birds may occupy the same cavity in consecutive years (Hayward 1993).

During the nesting season, the male Boreal Owl seldom roosts close to the nest itself. In fact, Hayward (1993) found roosting male owls anywhere from 50 yards to roughly 570 yards (15.24m to 174 m.) from the nest.

In a captive breeding situation (McKeever per.com.), when the male of the breeding pair died, the female continued to care for the chicks and a new male replaced the male that had passed away. Unfortunately, I do not know whether or not this type of activity occurs in the wild.

Nest Heights

Tufts (1925) and Prebles (1908) found the birds nesting at heights of 10 feet (3.04m) and 20 feet (6.09m) above ground. Lawrence (1932) found a nest in Winnipeg that was 18 feet (5.48 m) above ground. According to Hayward (1993), the owls nest relatively high ranging from 19.68 ft. to 82 ft. (6 to 25 meters) from the ground, with the cavities averaging at least 51% of the tree height. Bondrup-Nielsen (1978) found cavities ranging from 26 feet to almost 56 feet (11-17 m) from the ground. As with Northern Saw-whet Owls, Boreal Owls tend to occupy one of the highest nests when the tree has more than one cavity in it.

The size of the nest entrance varies quite a bit. It can be as small as a flicker-sized entrance (about three inches (76 cm)) to as large as eight inches (23 cm) and even larger. Bondrup-Nielsen (1978) found the entrance holes to the bird's cavities ranged from two-and-a-quarter inches by two-and-a-quarter inches to five-and-a-half inches by two-and-three-quarters inches (6 x6 cm to 14 cm x7 cm). Also the depth of the cavities was as deep as 13 5/8 inches (35 cm).

Eggs and Incubation

The clutch size of the Boreal Owl is between three and seven eggs, with the average being between four and six eggs. They average 32.3 by 26.9 millimeters, are pure white in color, and oval in shape (Bent 1939). The location of the nest seems to have some bearing on the time of egg laying. The female lays her eggs between late March and May, with incubation being an average of 29 days (Hayward 1993). Egg laying dates in Colorado (Palmer 1986) were from 17 April to 1 June. She begins egg laying from one to 19 days after the initiation of the occupation of her nest (Hayward 1993).

The female incubates exclusively, while the male feeds her throughout the evening. She will leave the nest each night for her personal needs, although it is only for a short time, especially when the chicks are very young. For the first three weeks after the young hatch, the male delivers food to the female, who in turn feeds the young (Hayward 1993).

During incubation and feeding the young for the first two to three weeks, the female relies completely on her mate for her own food and the food for her chicks (Bondrup-Nielsen 1978).

The Young

The owlets hatch in the order that the eggs were laid and, at hatching, the young are about nine grams (Hayward 1993). In 1978, Bondrup-Nielsen found the hatching dates of three eggs in the same nest being 23 May, 26 May, and 27 May respectively. Holt and Ermatinger (1989) saw four owlets fledge the nest between 20 and 24 June 1988.

As with most, if not all, owlets, young Boreal Owls are born with eyes closed and their bodies covered with white down. At four days old, dark pinfeathers begin appearing. At seven days, the flight feathers begin appearing and by the 11th day, the owls' eyes open. By the 25th day, the white spotting on crown and tail can be seen and by the 28th day, the young begin taking short flights. After two weeks the owlets can consume prey whole (Hayward 1993).

Young Boreal Owls leave home for the first time early in the evening, as do other nocturnal owls. At fledging, they are a chocolate-brown overall with bright yellow eyes and a pale bill. Between the eyes, the birds have a feathered white "X" which extends above the eyes and below the bill.

The forehead has a few white spots, as does the back and wings. The breast and belly are mostly the same chocolate-brown color, with some white blotching. The legs and feet are a bit lighter than the rest of the body and the talons are black. I am unaware of the time that it takes the young birds to molt into their adult plumage. However, I would suspect that it probably occurs over a 30-day period between mid-July and late August.

While in the nest, the nestlings give a *Chirp* call (Bondrup-Nielsen 1978), which prior to and after fledging apparently motivates the female to feed her chicks. As the chicks leave the nest, they often utter a *Chatter* call, which is a grouping of notes uttered in rapid succession. The latter was heard uttered as a researcher was removing the nestlings from the nest and again as they were trying to get under their mother, presumably to be protected.

Over the years researchers have captured and banded several nestlings and fledglings. The weight of the owlets ranged from 4.23 oz. (120 grams) for the males to 5.32 oz (151 grams) for the females.

After talking to three different Boreal Owl researchers, I learned that just before fledging, a percentage of females leave their nest area and the males have to provide for and protect the fledglings. Personally, I can't quite understand why this occurs and the other researchers didn't either.

Fledgling Boreal Owl.

Juvenile Boreal Owl. © Bill Schmoker

Small Mountain Owls · Part Four

Like other forest dwelling owls, the Boreal Owl often perches within the confines of the forest.
© Bill Schmoker

The Boreal Owl   165

Chapter Four

Hunting and Diet

Hunting primarily after dark, the Boreal Owl is a kind of sit and wait hunter, often waiting 10 minutes or more before making an attempt at a kill. The few times I watched the species hunting, the birds would wait, perching roughly 10 to 15 feet (3 to 4.57 meters) from the ground for just a few minutes before flying to another perch in search of an unsuspecting rodent. John Rawinski and a few other researchers placed a mouse near a Boreal Owl and as soon as they moved from sight, the owl grasped the mouse (Rawinski per. com.)

Boreal Owls often perch for several minutes before progressing to another perch, looking and listening for an unsuspecting victim to capture. When moving through the woods in search of prey, they often fly in a zigzag pattern, taking short flights between perches (Hayward 1993). During winter, when the snow becomes deep, the owls will wait until a Red-backed Vole, Western Jumping Mouse or other small mammal moves out from under the snow into the tree where it can be captured.

Another hunting technique employed by this species, especially when the snow is deep, is to sit close to the ground watching the area around the tree trunks in hopes that a mouse or vole comes to the base of the tree searching for seeds or other item to consume, at which time the owl can capture it without having to plunge into the snow to make a capture.

Hayward et al. (1993) observed one of these owls grab an unsuspecting songbird one evening. They were observing an owl as it began surveying a tight clump of Lodgepole Pine branches. After 12 minutes it flew into the tree to capture a roosting songbird. Another owl, hunting by day, spent 10 minutes watching some tall grasses before flying about 13 feet (four m) to capture a Red-backed Vole deep within the grass.

Ryder, Palmer, and Rawinski (1987) watched Boreal Owls hunting on twenty-seven occasions. They watched the owls plunge into small shrubs to capturing voles and also noted that moving prey was captured more often than inactive prey.

The primary diet of this species seems to be small to medium-sized mammals, including mice, voles, shrews, etc.

The prey species that Ryder et al. (1987) documented in Colorado consisted of Deer Mice, Southern Red-backed Voles, Montane Voles, a Least Chipmunk, a Long-tailed Vole, and a shrew, which species was unidentified. Birds that were identified included an American Robin, a Dark-eyed Junco, and a Mountain Chickadee.

Other species represented in the diet of the Boreal Owl include Short-tailed Shrews, Northern Pocket Gophers, White-footed Mice, Meadow Voles, Northern Bog Lemmings, and Northern Flying Squirrels.

Boreal Owl with
Red-backed Vole.

Both Ryder et al. (1987) and Hayward and Hayward (1993) found that over half of the prey items found in both studies were voles and conversely birds made up a relatively small portion of the species' diet.

Other birds documented as food items for the owl include Gray Jay, Hermit Thrush, Pine Siskin, Red Crossbill, Common Redpoll, warblers, and kinglets, along with woodpeckers (all from Hayward 1993).

Pine Grosbeaks and other forest songbirds can be part of the Boreal Owl's diet.

Food Caches and Pellets

Before and during nesting, Boreal Owls often cached prey in and near the nest cavity, where both adults can retrieve a morsel when needed. The food cache outside the nest is most often in the fork of a tree branch or on the needles of a spruce or fir near the nest (Hayward 1989).

Most owl species usually have a roost tree, under which you can find pellets that the birds have coughed up. Like all owl species, Boreal Owls do regurgitate pellets; however, finding them can be a bit challenging because the birds seldom, if ever, roost in the same tree from one day to the next.

Boreal Owls, most often, cough up at least one pellet per day that is roughly 37 x 13.37 mm. Unlike saw-whet owls, the pellets of Boreal Owls often contain the entire skull of the species they've consumed. They often swallow the head of their prey intact and subsequently the skull can be found complete with the jaws in place as they would be if someone had placed them together. They rarely however, consume an item whole, and more often consume their prey in sections.

Trapping Boreal Owls

When trying to capture these owls, mist nets and a captive mouse or vole seem to be the preferred technique. Ryder et al. (1987) placed two 40-foot (12 m) by eight-and-a-half-foot (2.6 m) mist nets (large nylon mesh nets with four pockets that stretch the length of the net) in a "V" with a bal-cha-tri (small wire cage covered in nylon nooses) with a live mouse or vole placed inside. The nets and mouse are then placed in a probable flight path of the owl.

Then the bird's primary call is broadcast to attract it to the area. The bird either flies into the area and hits the mist net or the noose cage and gets captured by entangling its feet in the nooses. Either way, the bird can then be extracted, measured, aged, banded, and the individual's sex determined, before the bird is released.

Hayward et al. (1993) used a similar technique to capture owls. They were trapping birds during winter as well as spring. After locating singing males, mist nets were placed along the flyways that the birds use. Bal-cha-tri traps loaded with mice were placed near the mist nets and the staccato call is played to attract the birds. As the birds moved through the area in search of what they believe to be a calling male, they eventually fly into the mist net and fall into one of the pockets and are captured. Either way, the information taken from the bird is the same.

Chapter Five

Roosting Behavior

Most of the owl species that I've studied tend to roost in the same general area each day and sometime even in the same tree and to a lesser extent on the same branch from day to day. Boreal owls seem to be quite different and in fact seem to roost wherever they end up before sunrise.

Boreal Owls rarely roost in the same general area or even the same stand of trees from day to day. Hayward (1993) found 14 owls roosting on 159 occasions and the birds only roosted in the same tree in eight cases.

They also found that the birds tended to roost in the lower sections of the trees more often in winter than summer and subsequently roosted higher in the trees during summer. Also of note, the birds tended to roost in more dense cover during summer than winter, which may be due in part to the greater number of songbirds in their territories in the summer and the owls not wanting to be disturbed. Therefore, being better concealed might be to the owl's advantage.

During the day, the owls spend the majority of their time perched with either eyes closed or just barely open watching their surroundings. They sleep during the day but rarely sleep for more than 40 minutes at a time (Hayward et. al 1993). Prior to leaving their perch for their evening hunt, they will spend a bit of time preening.

Winter Boreal Owls

Boreal Owls seem to be an irruptive species having, in some years, numerous individuals showing up south of their normal wintering range (Eckert 1982). When this occurs, it most often means that the owls can't locate enough food within their normal wintering grounds so they move south in search of provisions.

Over the years, there have been many of these irruptions, with the most recent being during the winter of 2004-2005, when hundreds of Boreal, Hawk, and Great Gray Owls were seen in Minnesota and Wisconsin. The species congregated in those regions because of the large number of rodents present.

A similar event happened between November 1981 and April 1982, when 39 migrating and wintering Boreal Owls were found in Minnesota (Eckert 1982). These irruptions occur whenever the winter snows become too deep for the birds to locate and capture the prey they need.

Christmas Bird Counts

Since the first Christmas Bird Count in 1900 the Boreal Owl has been seen on only 34 different counts nationwide. The first was during the winter of 1961-62 and from then on there have been a handful of sightings. The highest numbers of individuals were counted during the 2003-04 count with 24 Boreal Owls seen in eight different count circles.

Throughout the United States, the highest number of owls seen was on the 101st count, which was in 2000-2001 with ten birds having been seen in five different circles.

Longevity and Mortality

The oldest wild Northern Saw-whet Owl, as of 2004, was more than 10 years. I would presume that the oldest wild Boreal Owl is at least that old, if not older. However, the Bird Banding Laboratory has no longevity record for a wild Boreal Owl. Kay McKeever, president of the Owl Foundation in Ontario, Canada, told me that she had two Boreal Owls, one male and one female that lived 18 and 16 years respectively.

As with many other North American birds, the Boreal Owl at one time was very susceptible to the West Nile Virus, which did account for a number of Boreal Owl deaths in the past.

During the winter of 1981-82, and again in 2004-2005, several of these owls were found dead in northern Minnesota and northern Wisconsin. The cause of death appeared to be starvation. Apparently the snow was so deep that the birds couldn't get to the small mammals underneath it. In 2004-05, over 150 Boreal Owls were found dead due to starvation in those areas.

In 1963, biologist J.R. Koplin shot and killed a juvenile female owl in Colorado. This type of activity was more widely practiced in the early 1900s and is prohibited by law now.

Animals such as Red Squirrels and Pine Martens are responsible for a portion of the deaths of nestling, fledglings, and females on the nest (Hayward and Hayward 1993). Diurnal raptors such as Northern Goshawks, Cooper's Hawks, and Great Horned Owls will on occasion prey upon both adult and juvenile owls as well.

At least two of the owls were found dead, being inadvertently caught in mammal traps. One bird had a crushed skull and the other was caught on its left side (Siddle 1984). There are a few accounts that may suggest that some Boreal Owls may even resort to cannibalism in the winter when food is scarce.

Above:
Northern Pygmy-Owl fledgling in the author's hand.
© Jim Osterberg

Left:
Flammulated Owl on the bottom, Boreal Owl in the middle, and Northern Saw-whet Owl on the top. All live together in harmony at the Birds of Prey Foundation in Colorado.

Scientific Names of Birds, Plants, and Animals

Birds:

American Crow (*Corvus brachyrhynchos*)
American Kestrel (*Falco sparverius*)
American Robin (*Turdus migratorius*)
Bald Eagle (*Haliaeetus leucocephalus*)
Band-tailed Pigeon (*Columba fasciata*)
Barn Owl (*Tyto alba*)
Barred Owl (*Strix varia*)
Black-billed Magpie (*Pica pica*)
Black-capped Chickadee (*Pocile atricapilla*)
Bonaparte's Gull (*Larus philadelphia*)
Boreal Owl (*Aegolius funereus*)
Brown Creeper (*Certhia americana*)
Broad-tailed Hummingbird (*Selasphorus platycercus*)
Brown-capped Rosy-Finch (*Leucosticte australis*)
Budgerigar (*Melopsittacus undulatus*)
Burrowing Owl (*Athene cunicularia*)
California Quail (*Callipepla californica*)
Cassin's Finch (*Carpodacus cassinii*)
Chipping Sparrow (*Spizella passerina*)
Common Grackle (*Quiscalus quiscula*)
Common Nighthawk (*Chordeiles minor*)
Common Poorwill (*Phalaenoptilus nuttallii*)
Common Raven (*Corvus corax*)
Common Redpoll (*Carduelis flammeus*)
Common Yellowthroat (*Geothlypis trichas*)
Cooper's Hawk (*Accipiter cooperii*)
Dark-eyed Juncos (*Junco hyemalis*)
Downy Woodpecker (*Picoides pubescens*)
Dusky Grouse (*Dendragapus obscurus*)
Eastern Meadowlark (*Sturnella magma*)
Eastern Screech Owl (*Megascops asio*)
Elf Owl (*Micrathene whitneyi*)
European Pygmy-Owl, (*Glaucidium passerinum*)
European Starling (*Sturnus vulgaris*)
Ferruginous Pygmy-Owl (*Glaucidium brasilianum*)
Flammulated Owl (*Otus flammeolus*)
Franklin's Gull (*Larus pipixcan*)
Golden-crowned Kinglet (*Regulus satrapa*)
Great Gray Owl (*Strix nebulosa*)
Great Horned (*Bubo virginiianus*)
Green-tailed Towhees (*Pipilo chlorurus*)
Hairy Woodpecker (*Picoides villosus*)
Harris's Hawk (*Parabuteo unicinctus*)
Harris's Sparrow (*Zonotrichia querula*)
Hermit Thrush (*Catharus guttatus*)
House Sparrow (*Passer domesticus*)

House Wren (*Troglodytes aedon*)
Lesser Goldfinch (*Carduelis psaltria*)
Long-eared Owl (*Asio otus*)
MacGillivray's Warbler (*Oporonis tolmiei*)
Mountain Bluebirds (*Sialia currucoides*)
Mountain Chickadee (*Poecile gambeli*)
Mourning Dove (*Zenaida macroura*)
Northern Cardinal (*Cardinalis cardinalis*)
Northern Flicker (*Colaptes auratus*)
Northern Goshawk (*Accipiter gentilis*)
Northern Mockingbird (*Mimus polyglottos*)
Northern Saw-whet Owl (*Aegolius acadicus*)
Northern Shrike (*Lanius excubitor*)
Peregrine Falcon (*Falco peregrinus*)
Pileated Woodpecker (*Dryocopus pileatus*)
Pine Grosbeak (*Pinicola enucleator*)
Pine Siskin (*Carduelis pinus*)
Pygmy Nuthatch (*Sitta pygmaea*)
Red-breasted Nuthatch (*Sitta canadensis*)
Red Crossbill (*Loxia curvitostra*)
Red-napped Sapsucker (*Sphyrapicus nuchalis*)
Red-tailed Hawk (*Buteo jamaicensis*)
Red-winged Blackbird (*Agelaius phoeniceus*)
Rosy Finch (*Leucosticte species*)
Ruby-crowned Kinglet (*Regulus calendula*)
Ruffed Grouse (*Bonasa umbellus*)
Sharp-shinned Hawk (*Accipiter striatus*)
Short-eared Owl (*Asio flammeus*)
Spotted Owl (*Strix occidentalis*)
Spotted Sandpiper (*Actitis macularia*)
Steller's Jay (*Cyanocitta stelleri*)
Swainson's Thrush (*Catharus ustulatus*)
Swamp Sparrow (*Melospiza georgiana*)
Three-toed Woodpecker (*Picoides tridactylus*)
Townsend's Solitaire (*Myadestes townsendi*)
White-breasted Nuthatch (*Sitta carolinensis*)
White-crowned Sparrow (*Zonotrichia leucophrys*)
Williamson's Sapsucker (*Sphyrapicus thyroideus*)
Wilson's Snipe (*Gallinago delecta*)
Yellow-rumped Warbler (*Dendroica coronata*)

Trees and Shrubs:

Balsam Poplar (*Populus balsamifera*)
Black Spruce (*Picea mariana*)
California Black Oak (*Quercus kelloggii*)
Common Juniper (*Juniperus communis*)
Cottonwood (*Populus species*)
Douglas fir (*Pseudotsuga menziesii*)
Engelmann Spruce (*Picea engelmanni*)
Rocky Mountain Juniper (*Juniperus scopulorum*)
Lodgepole Pine (*Pinus contorta latifolia*)
Mountain Alder (*Alnus tenuifolia*)
Oak (*Quercus species*)
Paper Birch (*Betula papyrifera*)
Ponderosa Pine (*Pinus ponderosa*)
Poplar (*Populus species*)
Quaking Aspen (*Populus tremuloides*)
Subalpine fir (*Abies lasiocarpa*)
Arizona Sycamore (*Platanus wrightii*)
Wax Currant (*Ribes cereum*)
White Cedar (*Thuja occidentalis*)
White Spruce (*Picea glauca*)

Animals:

Bobcats (*Lynx rufus*)
Deer Mouse (*Peromyscus maniculatus*)
Elk (*Cervus elaphus*)
Golden-mantled Ground Squirrel
   (*Spermophilus lateralis*)
Gray Jay (*Perisoreus canadensis*)
Hispid Cotton Rat (*Sigmodon hispidus*)
House Mouse (*Mus musculus*)
Harvest Mouse (*Reithrodontomys species*)
Least Chipmunk (*Eutamias minimus*)
Long-tailed Vole (*Microtus longicaudus*)
Mountain Lion (*Puma concolor*)
Meadow Vole (*Microtus pennsylvanicus*)
Montane Vole (*Microtus montanus*)
Northern Bog Lemming (*Synaptomys borealis*)
Northern Flying Squirrel (*Glycomys sabrinus*)
Northern Pocket Gopher (*Thomomys talpoides*)
Northern Pygmy-Mouse (*Baiomys taylori*)
Pine Marten (*Martes americana*)
Wood Rat (*Neotoma cinerea*)
Red Squirrel (*Tamiasciurus hudsonicus*)
Raccoon (*Procyon lotor*)

Smokey Shrew (*Sorex fumeus*)
Short-tailed Shrew (*Blarina brevicauda*)
Southern Red-backed Vole (*Clethrionomys gapperi*)
Shrew (*Sorax species*)
Texas Kangaroo Rat (*Dipodomys* species)
Western Jumping Mouse (*Zapus princeps*)
Western Red-backed Vole (*Myodes californicus*)
Weasel (*Mustela species*)
White-footed Mouse (*Peromyscus leucopus*)
Woodland Mouse (*Peromyscus leucopus*)

Lizards and Snakes:

Six-Lined Racerunner (*Cnemidophorus sexlineatus*)
Texas Horned Lizard (Phrynosoma cornutum)

Moths:

Columbia Silk Moth (*Hyalophora columbia*)

References Cited

Altmann, S.A. 1956. Avian mobbing behavior and predator recognition. *Condor* 58: 241-253.

American Ornithologists Union. 2004 Checklist of North American Birds 7th edition. Am. Ornithol. Union, Washington D.C.

Bailey, A.M. and R.J. Neidrach. 1965. Pictorial Checklist of Colorado Birds. Denver Museum of Natural History, pp. 426-427.

Balda, P.B. and B. C. Mcknight and C.D. Johnson. 1975. Flammulated Owl Migration in Southwestern United States. *Wilson Bull.* Vol. 87, No. 4, pp.520-533.

Baldwin, P.H. and J.R. Koplin. 1966 The Boreal Owl as a Pleistocene Relict in Colorado. *Condor* 68:299-300.

Balgooyen, T.G. 1969. Pygmy Owl Attacking California Quail. *Auk* 86: 358.

Bent A.C. 1938. *Life Histories of North American Birds of Prey Part 2.*

Bishop, L.B. 1931. Sexual dichromatism in the Pygmy Owl. *Proceedings of the Biological Society of Washington.* 44: 97-98.

Bloom, P.H. 1983. Notes on the Distribution and Biology of the Flammulated Owl in California. *Western Birds* 14: pp.49-52.

Bondrup-Nielson, S. 1978. *Vocalizations, Nesting and Habitat Preferences of the Boreal Owl (Aegolius funereus) in North America.* Thesis pp. 1-158.

Bondrup-Nielson, S. 1984. Vocalization of the Boreal Owl. *Aeoglius funereus richardsoni* in North America. *Canadian Field-Naturalist* 98(2): 191-197.

Borell, A.E. 1937. Cooper's Hawk eats Flammulated Screech Owl. *Condor* 39:44.

Bouricius, S. 1987. An observation of Northern Pygmy Owl predation upon Harris's Sparrow. *C.F.O. Journal* 21: pp 18-21.

Bull, E. L. and R.G. Anderson. 1978. Notes on Flammulated owls in Northeastern Oregon. *Murrelet* 59: 26-28.

Brock, E.M. 1958. Some prey of the pygmy owl. *Condor* 60: 338.

Bull, E.L., J.E. Hohmann, and M.G. Henjum. 1987. Northern Pygmy-Owl Nests in Northern Oregon. *The Journal of Raptor Research Foundation* 21: 77-78.

Castle, G.B. 1937. The Rocky Mountain Pygmy Owl in Montana. *Condor* 39: 132.

Cartron, J.L. E. and D.M. Finch. 2000. *Ecology and Conservation of the ferruginous Pygmy Owl in Arizona.* USDA.

Catling. P.M. 1971. A behavioral attitude of Saw-whet and Boreal Owls. *Auk* 89, pp. 194-195.

Cous, E. 1882. *The Cous Checklist of North American Birds.* 2nd Edition. Boston: Estes and Laurat.

Clabaugh, E.D. 1933. Food of the Pygmy Owl and Goshawk. *Condor* 35: 80.

Daggett, F. S. 1913. Another instance of Cannibalism in the Spotted Owl. *Condor* 15: 40-41.

De Latorre, J. 1990. Owls: *Their Life and Behavior.* Crown Publishers, Inc. New York.

DeLong, J.P. Fall 2000. *Flammulated Owl banding study in the Manzo Mountains of Central New Mexico.* pp.1-10.

DeLong, J.P. 2003. *Flammulated Owl Migration Project Manzo Mountains New Mexico 2003* Report. pp. 1-10.

Deppe, C. et al. 2003. Effects of Northern Pygmy Owl (*Glaucidium gnoma*) eyespots on avian mobbing. *Auk* 120: 765-771.

Duncan, J.R. 2003. *Owls of the World. Their lives, Behavior and Survival.* Firefly Books.

Earhart, C.M. and N.K. Johnson. 1970. Size dimorphism and food habits of North American owls. *Condor* 72: 251-264.

Eckert. K.R. 1982. An invasion of Boreal Owls. *The Loon* Vol. 54. pp. 176-177.

Ely,D.C. 1992. *Proposal to Determine Nesting Density Differences of the Northern Pygmy-Owl in 3 Major Colorado Forest Types.*

Evanich, J.E. *The Birder's Guide to Oregon.* Portland Audubon Society.

Furniss, R. L. And V. M. Lucas. 1970. *Western Forest Insects.* USDA Forest Pacific Service, Pacific Northwest Forest and Range Expt. Sta. Misc. Publication 1339: 1-654.

Geise, A.R. 1999. *Habit selection by Northern Pygmy Owls on the Olympic Penninsula,* WA. M.S. thesis, Oregon State, Corvallis.

Gullion, G. and T. Martinson, 1984. Grouse of the North Shore. pp. 17.

Hannah, K.C. 1999. Status of the Northern Pygmy Owl (*Glaucidium gnoma californicum*) in Alberta. *Alberta Wildlife Status Report* no. 20. Fish and Wildlife Division, Alberta Environmental Protection, Edmonton.

Hasenyager, R.N., J.C. Peterson, and A.W. Heggen. 1979. Flammulated Owl nesting in a Squirrel Box. *Western Birds* 10: 224. 1979.

Hayward, G.D. and E.O. Garton, 1983. First nesting record for Boreal Owls in Idaho, short communications. *Condor* 85:501.

Hayward, G.D., P.H. Hayward, and E.O. Garton, 1987. Revised Breeding Distribution of Boreal Owl in the Northern Rocky Mountains. *Condor* 89:431-432. The Coopers Ornithological Society 1987.

Hayward, G.D. and P.H Hayward. 1993. The Birds of North America, No. 63.

Hayward, G.D. and P.H. Hayward and E.O. Garton 1993. Ecology of Boreal Owls in the Northern Rocky Mountains, USA. *Wildlife Monographs* 124, 1-59.

Holman, F.C. 1926. Nesting of the California Pygmy Owl in Yosemite. *Condor* 28: 92-93

Holt, D.W. and J.L. Peterson, 2000. *The Birds of North America*, No. 494. Northern Pygmy Owl.

Holt, D.W. and R. Kline. 1989. Glaring gnome. *Montana Outdoors* 20: 13-15.

Holt, D.W. and D. Ermatiner 1989. First confirmed nest site of Boreal Owls in Montana.

Holt, D.W. and L.A. Leroux. 1996. Diets of Northern Pygmy Owls and Northern Saw-whet Owls in west-central Montana. *Wilson Bull.* 108: 123-128.

Holt, D.W. and W.D. Norton. 1986. Observations of nesting Northern Pygmy Owls. *J. Raptor Res.* 20: 39-41.

Holt, D.W., R. Kline, and L. Sullivan-Holt. 1990. A description of Atufts@ and concealing posture in Northern Pygmy Owls. *J. Raptor Res.* 24: 59-63.

Howell, S.N.G. and S. Webb. 1995. *The Birds of Mexico and Northern Central America.* Oxford University Press, New York.

Hume, R. 1991. *Owls of the World.* pp.96-101.

Johnsgard, P.J. 1988. North American Owls: Biology and Natural History. Smithsonian Institution. Washington D.C.

Johnsgard, P.J. 2002. *North American Owls: Second Edition: Biology and Natural History.* Smithsonian Institution. Washington D.C.

Johnson, N.K., and W.C. Russell. 1962. Distributional data on certain owls in the western Great Basin. *Condor* 64: 513-514.

Jones, S. 1991. Distribution of small forest owls in Boulder County Colorado. *C.F.O. Journal* 25: 55-70

Jones, S.E. 1998. Northern Pygmy-Owl. pp. 218-219 of *Colorado Breeding Bird* (H.E. Kingery, ed.), Colorado Bird Atlas Partnership and Colorado Division of Wildlife, Denver, Colorado.

Keith, L.B. 1963. *Wildlife's Ten Year Cycle.* University of Wisconsin Press, Madison. WI.

Kellomaki, E. 1977. Food of the Pygmy Owl (*Glaucidium passerinum*) in the breeding season. *Ornis Fennica.* 54: 1-29.

Kendeigh, S.C. 1969. Tolerance of cold and Bergmanns rule. *Auk* 86 (1): 13-25.

Kendeigh, S.C. 1970. Energy requirements for existence in relation to body. *Condor* 72: 60-65.

Kenyon, K.W. 1946. Cause of Death of a Flammulated Owl. The Condor. *Condor* 49. 88.

Kimball, H.H. 1925. Pygmy Owl killing a quail. *Condor* 27: 209-210.

Kleinschnitz, F.C. 1939. Field Manual of Birds Rocky Mountain National Park.

Kullberg, C. 1995. Strategy of Pygmy Owl while hunting avian and Mammalian prey. *Ornis Fenn.* 72: 72-78.

Ligon, J.D. 1969. Some aspects of temperature relations in small owls. *Auk* 86: 458-472.

Linkhart, B.D., E.M Evers, J.D. Megler, E.C. Palm, C.M. Salipante, and S.W. Yanco. 2008. First Observed Instance of Polygyny in Flammulated Owls. *The Wilson Journal of Ornithology* 120(3): 645-648. 2008.

Linkhart, B.D., R.T Reynolds. 2007. Return Rate, Fidelity, and Dispersal in Breeding Population of Flammulated Owls (*Otus flammeolus*). *The Auk* 124 (1) 264-275.

Linkhart, B.D., R.T Reynolds. 2004. Longevity of Flammulated Owls: additional records and comparisons to other North American strigiforms. *J.Field Ornithol.* 75 (2): 192-195.

Linkhart, B.D., R.T. Reynolds, and R.A. Ryder, 1998. Home range and Habitat of Breading Flammulated Owls in Colorado. *Wilson Bull.* 110 (3), 1998 pp. 342-351.

Linkhart, B.D., R.T Reynolds. 1997. Territories of Flammulated Owls (Otus flammeolus) Is Occupancy a Measure of Habitat Quality? Biology and Conservation of Owls of the Northern Hemisphere. Second International Symposium. pp. 250-254.

Linkhart, B.D. 1984. *Range, activity and habitat use by Flammulated Owls in a Colorado Ponderosa Pine forest*. M.S. thesis, Colorado State University, Fort Collins Colorado.

Mannan, R.W., C.W. Boal. 1990. Goshawk diets in logged and unlogged ponderosa pine forest in Arizona. Progress report, challenge cost-share agreement, University of Arizona and Kaibab National Forest. pp.10.

Marshall, J.T. 1939. Territorial behavior of the Flammulated Screech Owl. *Condor* Vol. XLI, pp.71-78.

Matthiae, T.M. 1982. A Nesting Boreal Owl in Minnesota. *The Loon* Vol. 54, pp. 212-213.

McCallum, D.A. 1994. Review of Technical Knowledge: Flammulated Owls. In *Flammulated, Boreal and Great Gray Owls in the United States*, pp. 14-41.

McCallum, D.A. and F.R. Gehlbach, 1988. Nest-site preferences of Flammulated Owls in Western New Mexico. *Condor* 90: 653-661.

Meehan, R.H. 1980. *Behavioral significance of Boreal Owl vocalization during breeding season*. M.Sc. thesis, Univ. Alaska Fairbanks.

Meehan R.H. and R.J. Ritchie 1982. Habitat requirements of Boreal and Hawk Owls in Interior Alaska. Proceedings of a symposium and workshop Raptor Management and Biology in Alaska and Western Canada. pp. 188-195.

Merle, R.L., L.R. DeWeese, and R.E. Pillmore. 1980. Brief Observation on the Breeding Biology of the Flammulated Owl in Colorado. *Western Birds* 11: 35-46.

Merrill, J.C. 1888. Notes on the birds of Fort Klamath, Oregon. *Auk* 5: 139-146.

Michael, C.W. 1927. Pygmy Owl: The little demon. *Condor* 29: 161-162.

Miller, A.H. 1934. The vocal apparatus of some North American owls. *Condor* 36: 204-213.

Mysterud, I., and H. Dunker. 1979. Mammal ear mimicry; a hypothesis on the behavioral function of owls horns. *Anim. Behav.* 27: 315.

Norton, W.D., and D.W. Holt. 1982. Simultaneous nesting of Northern Pygmy Owls and Northern Saw-whets in the same sang. *Murrelet* 63: 94.

Norberg, A 1987. Evolution, structure, and ecology of northern forest owls. pp. 9-43 Biology and conservation of northern forest owls. Symposium proceedings. USDA forest and Range experimental station Ft. Collins, CO.

Norberg, A 1968. Physical factors in directional hearing in Aegolius funereus with special reference to the significance of the asymmetry of the external ears. *Auk. Zool.* Ser. 2, 20: pp. 181-204.

Packard, F.M. 1945. The Birds of Rocky Mountain National Park. *Auk* 62: 371-394.

Palmer, D.A. and R.A. Ryder. 1984. The first documented Boreal Owl in Colorado. *Condor* 86: 215-217.

Palmer, D.A.1986. *Habitat selection movements and activity of Boreal and Saw-whet Owls*. Thesis. CSU. Ft. Collins CO.

Palmer, D.A. 1987. Annual, Seasonal, and nightly variation in calling activity of Boreal and Northern Saw-whet Owls. pp 162-168.

Palmer, D.A. and J.J. Rawinski. 1988. A Technique for locating Boreal Owls in the fall in the Rocky Mountains. *Oregon Birds* 1491: 23.

Perrone, M.J. 1981. Adaptive significance of ear tufts in owls. *Condor* 83: 383-384.

Pierce, W.M. 1921. California Pygmy Owl from Cucamonga Canyon, southern California. *Condor* 23: 96.

Phillips, A.R. 1942. Notes on the migration of Elf and Flammulated screech Owls. *Wilson Bull.* 54: 132-137.

Rashid, S. 1999. Northern Pygmy Owls in Rocky Mountain National Park. C.F.O. Journal. 33: 94-101.

Rashid, S. 1999. The Forest Gnome. *Rocky Mountain Nature Association Quarterly* pp.4-13.

Rashid, S. 2005. Small Owls of Rocky Mountain National Park. *Colorado Birds. The Colorado Field Ornithologists Quarterly* pp. 17-23.

Rawinski, J.J., R.Sell, P. Metzger, H. Kingery, and U. Kingery. 1993. Young Boreal Owls found in the San Juan Mountains Colorado. *C.F.O. Journal* Vol. 27, No. 2, pp. 56-59.

Reynolds, R.T. et al. 1992. Management Recommendations for Northern Goshawk in the Southern United States. USDA. Forest Service. pp. 52.

Reynolds, R.T. et al. National Wildlife Federation Small Forest Owls Western Management symposium and workshop, pp. 138-139.

Reynolds, R.T. and B.D. Linkhart. 1987b. The nesting biology of the Flammulated Owls in Colorado. Pp. 239-248 in R.W. Nero, R.J. Clark, R.J. Knapton, and R.H. Hamre, editors, *Biology and Conservation of Northern Forest Owls*.

Richardson, C.H. 1906. Cannibalism in owls. *Condor* 8: 57.

Richmond, M.L., L.R. Deweese, and R.E. Pillmore. 1980. Brief observations on breeding biology of the Flammulated Owl in Colorado. *Western Birds* 11:35-46.

Righter, B. 1995. Description of a Northern Pygmy Owl vocalization from the southern Rocky Mountains. *C.F.O. Journal* 29: 21-23.

Ryder, R.A. and D.A. Palmer & J.J. Rawinski. 1987. Distribution of Boreal Owls in Colorado. pp. 169-174.

Sibley, D.A. 2000. National Audubon Society, The Sibley Guide to Birds.

Siddle, C. 1984. Raptor Mortality on Northeastern British Columbia Trap line. *Blue Jay* Sept. 42(3).

Small, A. 1998. *California Birds: Their Status and Distribution*. Rev. ed. Ibis Publication Company.

Solheim, R. 1984. Caching behavior, prey choice and surplus killing by Pygmy Owls (*Glaucidium passerinum*) during winter, a functional response of a generalist predator. *Ann. ool. Fennici.* 21: 301-308.

Snyder, N.F.R., and J.W. Wiley. 1976. Sexual dimorphism in hawks and owls of North America. *Ornithological Monogram* Number 12.

Stahlecker, D.W. and J.J. Rawinski, 1990. First Records of the Boreal Owl in New Mexico. *The Condor* Vol. 92(2), pp. 517-519.

Terres, J.K. 1982. *The Audubon Society Encyclopedia of North American Birds*.

Toops, C. 1990. *The Enchanted Owl*. Voyager Press, Stillwater, Minnesota.

Voous, K. H. 1988. *Owls of the Northern Hemisphere*. pp.138-151.

Walsh, P.J. 1990. Nest of Northern Pygmy Owl in southeast Alaska. *Northwest. Nat.* 71: 97.

Whitaker, J.O. 1980. *The Audubon Society Field Guide to North American Mammals*. Alfred A. Knopf. New York.